Benjamin Wills Newton

Aids to Prophetic Inquiry

Benjamin Wills Newton

Aids to Prophetic Inquiry

ISBN/EAN: 9783337035846

Printed in Europe, USA, Canada, Australia, Japan

Cover: Foto ©berggeist007 / pixelio.de

More available books at **www.hansebooks.com**

AIDS

TO

Prophetic Enquiry.

BY

BENJAMIN WILLS NEWTON.

Third Edition Considerably Enlarged.

LONDON:
HOULSTON AND SONS,
9, PATERNOSTER BUILDINGS.
1881.

CONTENTS.

CHAPTER I.
Introductory Observations 1

CHAPTER II.
No Poetic Exaggeration in Scripture 22

CHAPTER III.
Some Objections to the Doctrine of the Millennial Reign considered 42

CHAPTER IV.
On Zechariah xii. and xiii. 58

CHAPTER V.
On Zechariah xiv. 71

CHAPTER VI.
Futurity of the Manifestation of Antichrist—his connexion with Jerusalem 88

CHAPTER VII.
Passages of Scripture respecting Antichrist compared . 115

CHAPTER VIII.
Thoughts on Matthew xxiv. 127

CHAPTER IX.
On Luke xxi. 149

CHAPTER X.
Remarks on the Prophetic Statements of Mr. Fleming . 170

CHAPTER XI.
The Prophetic System of Mr. Elliott and Dr. Cumming considered 200

CHAPTER XII.
Scriptural Proof of the First Resurrection . . . 274

CHAPTER XIII.
Remarks on Bishop Wordsworth's Lectures on the Apocalypse. 310

CHAPTER XIV.
Remarks on Indefectibility 387

CHAPTER XV.
Present Tendencies 414

Appendix 446

PREFACE TO THE THIRD EDITION.

───◆───

THE first edition of this work was published in 1848. Each succeeding year has deepened my conviction of the certainty and momentous importance of the truths feebly and imperfectly treated of in it.

When the Eastern branch of the Roman Empire shall have been resuscitated, and · when Egypt, Assyria and Jerusalem, shall have become the sphere in which those habits of thought and action now cherished in the West shall have attained the matured form of their development, the end will be very nigh. The very persons, places, and circumstances of which the Prophets and the Revelation speak, will then be actually present before the eyes of men.

At present, however, we must not delude ourselves with the hope that much attention will be by any bestowed on these things.

It is a time of restless activity, when men, and even Christians, are absorbed in pursuing their own schemes. Impatient of direction and control, they like to act unrestrainedly; and are very indisposed to accept principles that would check many a cherished plan, and blight many a hope.

There is no part of God's Word that we are more prone to repudiate than His prophetic testimony. It reveals the extent of our departure from His ways; it declares His coming judgments; it reveals the manner in which His kingdom when it comes, will subvert the methods and constructions of men.

Nor is there any light that Satan dreads more than the light of Prophetic Truth. It is truly a light that shineth in a dark place. He has kindled many fires to illumine this darkness, and has deceived men into the belief that the fires thus kindled from the pit have been kindled of God. He fears the effect that would be produced by an awakenment to this dread delusion.

For sixteen hundred years and more, Christendom has been deceived into the belief that the present period, in which Satan as the κοσμοκρατωρ sways the nations, is the period when Christ and the saints are reigning. The Teachers of Christendom, age after age, have taught that this is so. It is a doctrine so contrary to all that fact and conscience teach, that it has already prepared many hearts for Scepticism, and will yet prepare many more. Yet it is a doctrine that has by no means ceased to operate, and is by many, who hold high and influential positions, still vigorously maintained.

The danger of this doctrine is by few adequately apprehended. I have, therefore, at the suggestion of a valued friend, added to this volume some remarks on Lectures on the Apocalypse delivered by the present Bishop of Lincoln, before the University of Cambridge. His doctrine respecting the reign of Christ and of His saints is not peculiar. It has been the standard doctrine of Christendom from the second century to the present hour. They who have been delivered from it, and who, when freed, have turned not to Scepticism but to the Bible, cannot be too thankful for the mercy of their deliverance. Statesmen, as well as ecclesiastics, have been for ages a prey to the deluding notion that there is, at present in the earth, a body of which it can be said that its "foundation is in the holy mountains," and that this body is appointed and commissioned by God to rule all nations. I have added some remarks on this subject also. Few are aware of the manner in which strong minds may become fascinated and spellbound by this thought; and how it has operated silently but potently in bringing about that condition of things in which England now finds itself involved.

There is not, I suppose, in Europe one God-fearing person who does not contemplate with awe the present condition of the nations. Never did men more need laws, government, and authoritative control, and yet never was self-will more insolently active. The authority, and often the existence of God, is being very extensively denied; and even our once favoured land has consented to say, that "civil duty should be separated from religious Truth." Our great legislative Assembly has branded that motto on its brow. Is not this an abandonment of God? What must be the result? Disorganisation and confusion—a confusion that would, if God were not to interfere, become the confusion of hell. Paris once afforded an example of this. Give to men increased freedom of action; puff them up with a vain conceit of their own dignity and power; destroy the distinctions which God has appointed in human society; honour and patronise those who take from men the fear of a judgment to come, and who say that the Bible is not to be confided in; that there is no certain Truth; that there is no eternal punishment; no Heaven; no Hell; no Devil; no God; and then trust to moral suasion and the natural intelligence and goodness and improveability of man's heart to supply the means whereby order shall be preserved, and government maintained—what would be the result? Demoniacal confusion.

It is true indeed that God will not wholly abandon the earth either to man or to Satan. He can preserve order by raising up rulers who will rule with a sceptre of iron, and scourge with scorpions. The Scripture speaks of the bonds of society becoming so relaxed that men are made like "fishes of the sea," like "creeping things that have no ruler over them," but it speaks also of a drag-net being prepared in which these fishes shall be caught and controlled; and the hand that will hold and use that net is the hand of that King for whom "Tophet is prepared." Is it not meet that those nations which have abandoned God should be delivered over to that King to whom the Dragon shall give his power, and throne, and great authority; and who will have power to crush, and will crush, all who refuse to honour and worship *him?*

We may despise the warnings of God, and fold our arms, and say complacently—"Magna est veritas et prævalebit." Truth will not prevail till that whirlwind, whereof God speaks, shall have gone forth and swept away every refuge of lies and every false hiding place. We have to watch now, not for the growth and progress of Truth, but for the growth and progress of Satan's last great lie.

The concluding chapters of the book which I preface with these remarks, touch upon these subjects in connexion with events which are now occurring amongst ourselves. The servants of God are not to be politicians. They are priests of God's sanctuary. But the priests' lips are to keep knowledge: they are to separate between the clean and the unclean, the precious and the vile. They are to watch the signs of the times; to see where evil and danger lurk, and to use every opportunity that occurs of warning all men, whether governors or slaves, princes or peasants. Paul's relation to Agrippa may guide us as to this. God sees a difference between governors and nations that "kiss," or do obeisance to the Son, and others who say of Jehovah and of Christ, "Let us break their bands asunder, and cast away their cords from us."

A time will come when some, whom God calls "understanding ones," will be raised up who shall speak of these things more wisely and more holily than any of us can speak of them now. Much darkness, and many sorrows will, I believe, overtake, and well nigh overwhelm the Church of God, before the time of that final testimony comes, for at present the Church of God is grievously turning away from the Word of God, and despising the very light which He has sent to guide them at this peculiar crisis of the Church's and the world's history. We never more needed holy separateness. "Be ye clean that bear the vessels of the Lord." "Touch not the unclean thing." I often think of the words, "It is of the Lord's mercies that we are not consumed, because His compassions fail not." We are little aware of the extent to which we have trifled with, or else ignored, the revealed truths of God's Word.

February, 1881.

Aids to Prophetic Enquiry.

CHAPTER I.

INTRODUCTORY OBSERVATIONS.

THERE are few periods in the history of the world that have been marked by deeper spiritual darkness than the commencing part of the eighteenth century. From 1700 to 1750, Protestantism seemed to have lapsed into lifeless formality—and this, together with the matured abominations of Popery, opened the way for that tide of infidelity, of which the French Revolution was the manifested result.

The latter part of the eighteenth century however, was, through the Lord's great mercy, marked by a very decided revival of evangelical truth. The effect of the writings and preaching of Whitfield, Romaine, Newton and others, was widely felt in Europe and in America; the simple message of salvation by grace through faith being received by many: and thus the very era in which Revolutionary Infidelity assumed its position of strength, was marked also by a renovation of truth, the effects of which will doubtless be discernible throughout the closing period of Antichristian evil.

The result of the evangelical testimony, to which I have referred, was seen in the commencement of the nineteenth century, in a widely-diffused desire to spread the knowledge of the Gospel. The Bible Society, as well as the chief Missionary Societies, date from the commencement of the present century, and in many instances were doubtless the result of a sincere desire, in the individuals who founded them, of promulgating truths which they had personally experienced to be precious. In order to stimulate missionary exertion, the Prophets of the Old Testament were appealed to. The time when creation shall cease from its groan, and the lion feed with the kid, and nations learn war no more, and the earth be full of the knowledge of the Lord, was said to be the sure result of that Gospel testimony which was being sent forth among the nations—a conclusion hastily indeed and erroneously formed; but God has mercy on ignorance when it is advancing out of darkness into light. More light therefore was given. The Prophetic Scriptures after having been once appealed to, were more continually used, and knowledge gradually increased.

The direction, however, of missionary effort towards THE JEWS, may be regarded as the chief cause of the advance made in prophetic truth. It would seem as though efforts spiritually to aid that people were peculiarly acceptable to God, even as He has said, "Blessed is he that blesseth thee." (Num. xxiv. 9.) The attempt to evangelize the Jews, necessitated a more frequent and more intelligent use of the *Old* Testament. The prospects of the Jews, as revealed in the Prophets, were more carefully enquired into; and it was soon discovered that on them and their repentance the blessing of the Earth at large was made instrumentally dependant: that it was therefore vain to expect the universal conversion of the nations, or the establishment of the reign of peace, until Israel should first "convert and be healed." It is the converted remnant of Jacob that is to be "in the midst of many peoples as a dew from the Lord, as the showers upon the grass, that tarrieth not for man, nor waiteth for the sons of men." (Mic. v. 7.) Ethiopia will not "stretch out her hands unto God," until Israel, long defiled, shall have become "as the wings of a dove covered with silver, and her feathers with yellow gold." God will *first* bless Israel, and *afterwards* "all the ends of the earth shall fear Him."

But even when this important truth had been recognised, it was still imagined that Israel was to be converted by the instrumentality of the Gentiles, from whom it was said the Jews *nationally* were to receive the Gospel—and a wrong translation of the 31st verse in the eleventh of Romans confirmed the delusion.* Gradually, however, it was discerned that the Gentile Churches were not to be made the means of converting the Jews—indeed that Israel, nationally, would, to the last, resist the Gospel as now preached in weakness, and would only bow to it and believe when suddenly confronted, like St. Paul in his blasphemy, by the visible glory of the Lord from Heaven. This indeed was an advance of unspeakable importance in the knowledge of Scripture. The Advent of the Lord Jesus in glory, was now recognised as the alone means appointed of God for the introduction of the Millennium; and it was clearly seen, that not only the conversion of Israel, but the binding of Satan, and the release of creation from its groan, were made dependent upon agency entirely different from any which God had connected with the suffering period of His Church's testimony. Indeed, no moral instrumentality such as the preaching of the Gospel, *could*

---

* The passage literally translated is as follows :—" For the gifts and calling of God are unrepented of. For as ye in times past disobeyed God, but now have obtained mercy by means of their disobedience : so these also have now disobeyed the mercy possessed by you, (*i.e.*, the Gospel,) that they also might become objects of mercy. For God hath shut up the whole in disobedience, that He might have mercy upon the whole."

either bind Satan, or raise the bodies of the saints who sleep. Both these things must be the result of *manifested* almighty power, and both, according to the arrangements of God, are made precursors of the Millennial reign.*

The *eleventh* and *fourteenth* chapters of Zechariah—the third of Joel, and the last of Isaiah, were among the passages appealed to, as clearly revealing that the manifestation of Christ in glory was to precede the conversion of Jerusalem, and the subsequent blessing of the Earth. The last chapter of Zechariah, after describing the day of the Lord's visitation on Jerusalem, says, "His feet shall stand in that day upon the mount of Olives; * * * * the Lord my God shall come, and all the saints with Thee; * * * * IN THAT DAY living waters shall go forth from Jerusalem, * * * and the Lord shall be king over all the Earth; * * * IN THAT DAY shall there be one Lord, and His name one." Joel speaking of the same period of

* Thus in the last verse of the 26th of Isaiah, it is said, "Behold the Lord *cometh out of His place* to punish the inhabitants of the Earth for their iniquity"—in the next verse Satan is described as smitten; after which it is said, "Israel shall blossom and bud, and fill the face of the world with fruit."

The same order of events is given in the nineteenth and twentieth chapters of Revelation; viz., Antichrist and the kings under him, smitten by Christ and His saints manifested in glory — Satan bound — Christ and the saints reigning.

visitation, says: "The sun and the moon shall be darkened, and the stars shall withdraw their shining. The Lord also shall roar out of Zion, and utter His voice from Jerusalem; and the heavens and the earth shall shake; but the Lord will be the hope of His people, and the strength of the children of Israel. So shall ye know that I am the Lord your God dwelling in Zion, My holy mountain: then shall Jerusalem be holy, and there shall no strangers pass through her any more." No one can doubt, after receiving the testimony of these passages, that the manifestation of the Lord in glory must *precede* the conversion of Jerusalem, and the reign of peace over the nations.*

* It should perhaps be mentioned, that just at this period, the progress of Prophetic Enquiry was greatly interrupted by the rise of Irvingism. Many who were involved in that heresy had become identified in general estimation with prophetic enquiry; and thus many were stumbled and driven back in alarm from the pursuit of a subject which seemed to involve such evil consequences.

They who teach on prophetic subjects cannot too carefully remember that the foundation truths of our holy faith (such, for example, as are given in the first eighteen Articles of the Church of England) are to be regarded as settled for ever. The Eternity and true Divinity of the Son of God—the doctrine of the Holy Trinity—the union of the Divine and human natures in the Person of Christ—the sinlessness and perfectness of all His experiences in the flesh—the complete and finished satisfaction made for all His believing people by His vicarious death upon the Cross, and the doctrine of justification by faith through grace, are holy and

The pre-millennial Advent of the Lord, when thus recognised and connected with the hopes which the Scriptures reveal as specially belonging to the

unalterable truths, the rejection of which unfits for further instruction in the word of God.

It should be remembered however, that Mr. Irving and those who followed him must on the whole be regarded as stern opposers of that particular use of Prophetic Truth which has since been found of such value. They were quite willing to condemn certain ecclesiastical systems to which they did not themselves belong; but they were *not* willing to let the light of prophecy fall upon the *secular* as well as the ecclesiastical systems around them. They were *not* willing that the nations governmentally should be viewed as " Beasts," and they shrunk from the practical consequences which such an acknowledgment would involve. Accordingly, few have done more than they to darken the testimonies both of Daniel and the Revelation.

In the earlier part of their career, they allowed that the millennial promises respecting Jerusalem and Israel were to be understood of Jerusalem and Israel, and that they were not to be appropriated to the Gentile Church, or any part of the Gentile Church, of the present dispensation; but as soon as they fell into the depth of their delusion, they forsook their former statements, and their pretended prophets began habitually to interpret Millennial Scripture of themselves—the object being to appropriate to themselves that place of supremacy in the earth which the Scripture reserves for converted Israel in the Millennium.

It is remarkable too, that Puseyism, which arose just at the same time as Irvingism, equally assailed the truth that "the gifts and calling of God are unrepented of"—and laboured to show that Israel nationally have forfeited their blessings for ever, and that consequently, those blessings

suffering family of faith, became, to many, the one absorbing thought in connection with prophecy; and they so lingered in it, that after a little, they contracted a dislike to all further prophetic investigation. They seemed to forget, that attention to signs predicted in the word of God is a duty; and that it is the prescribed means to that watchfulness which should characterise the servants of Christ at His appearing. Others, however, were led to continue their examination into prophetic Scripture, especially in connection with the prospects of Israel. It was soon found that there was a period in Israel's future history largely described in Scripture which had been entirely overlooked, even by those who had received the testimony to their Millennial glory, I mean, the period when they will again return in unbelief to their own Land, and there, under Antichrist, enter into their last great rebellion against God.

Antichristianism had been a subject on which

---

have been made over to the Gentile Church. These two most evil systems acted upon a different class of minds, but the minds on which they acted resembled one another, in this, that they were very proud, and anxious to gratify their pride by exalting themselves ecclesiastically. Hence the appropriation to themselves of the Millennial position of Israel. They wished "to reign as kings" before the time. They wished that "kings should be their nursing fathers, and their queens their nursing mothers." Romanism has ever coveted this.

the minds of Protestant Christians had often dwelt; but they had not connected it with Jerusalem, nor determined with sufficient care its characteristic features. In a word, they had not derived their thoughts respecting it simply from the Scripture. And even now, when the blasphemous infidelity of the French Revolution, as well as the more disguised Infidelity elsewhere apparent, has given living evidence that there may be something more terrible even than Popery itself, yet great reluctance has been shown to receive the history of Antichristianism as given in the word of God. Efforts are still being made to represent Antichristianism as *limited* to the comparatively narrow circle of Popery; and Daniel and the Revelation are still interpreted by many as if Jerusalem and Israel were entirely unconnected with the great consummation of evil in the latter day.

But as soon as it is distinctly seen in Daniel, that at the "time of the end" when the Divine indignation against Jerusalem is being "accomplished," there is a king to arise, who, doing in all things according to his will, is to glorify himself on the glorious holy mountain, Mount Zion; and is afterward to be destroyed at the time when Michael stands up to deliver Israel; it is evident, that if the prophetic parts of the New Testament, have been interpreted so as to omit this great concluding fact, such interpretation must have been emphatically wrong. The prophecies of the New

Testament *cannot* be interpreted in opposition to those of the Old. If the Old Testament has elaborately given the history of the last great Head of the Gentiles, and connected it with Jerusalem, its testimonies cannot be nullified by any thing revealed in the New Testament. And when we remember that the Disciples, to whom the prophecies of the New Testament were given, were Jews, that they were already acquainted with the Prophets, and that the prophecies of the New Testament are *professedly* supplemental to those already given in the Old (for Jesus came not to destroy the previous testimonies of God), we can easily see the importance of receiving the instructions of the Prophets, if we wish to apprehend the additional lessons of the Apostles. Accordingly, one of the most important consequences that has flowed from the examination of the Old Testament Scriptures is this, that it has necessitated a change in the habit of interpreting the Book of Revelation, because the system of interpretation that has in modern times prevailed, leaves no room for the accomplishment of the events declared in the Old Testament Prophets as yet to occur in Jerusalem.

Another consequence that followed the examination of the Old Testament Prophecies was the discovery that all historic detail* in Prophetic

---

* By "historic detail" is meant, all mention of localities, dates, personages, and the like. When Jerusalem ceased to have a *national* existence, all such detail was suspended.

Scripture is suspended, as soon as the narrative reaches the point at which Jerusalem ceases to hold a *national* existence in the Earth — that is, when it was desolated by the Romans, eighteen hundred years ago: and the detail is not resumed, until Israel again becomes possessed of a recognised national position; which will not be, until after they have returned to their own Land in unbelief, and have entered upon the last season of Antichristian abomination. Thus the many attempts that have been so laboriously made to find a thread of prophetic fulfilment throughout the past eighteen hundred years, have been found to be as wrong in principle as they have been unsatisfactory in result. The facts of prophetic history are made by Scripture to revolve around Jerusalem as their centre — and therefore any system of interpretation which violates this cardinal principle will soon find itself lost in inconsistency.

There has also been, not unfrequently, great inattention to the mode in which Prophetic Scripture, and indeed all Scripture is written. It is often read consecutively, as if the several chapters or visions, as they proceed, must necessarily follow each other *as to time*. It is forgotten that in all instruction, especially in narration, whenever the subject of instruction has various ramifications, it is necessary to recur several times to the same point, and to retrace; otherwise all the various features of the subject cannot be fully and distinctly given. Thus the first chapter of Genesis concludes with the day on which God rested from His work, when all was finished; but the second chapter, instead of being chronologically consecutive, returns to the period described in the preceding chapter, and narrates the manner of the creation of Eve. So also in Isaiah. The first chapter leads us through the whole scene of Israel's evil on to the hour of God's millennial interference, when "He will turn His hand upon them, and purely purge away their dross and take away all their tin, when He will restore their judges as at the first, and their counsellors as at the beginning." The subsequent chapters, therefore, necessarily *retrace*. So likewise in the Book of Daniel. The first vision, which is that of the Image, leads us on to the time when the Image is smitten—ground to powder, and the Stone that smites it becomes a great mountain and fills the whole Earth. No subsequent vision goes beyond this limit. They all retrace and develop other features belonging to the same period. The Book of Revelation is written on the same principle. The sixth chapter, which, speaking strictly, is the first of the series of prophetic visions, brings us to that final hour when men shall tremble and wail, because the great day of the Lamb's wrath will have come. The other visions on to the nineteenth chapter do but *retrace*, until the end of the nineteenth chapter describes the manifestation of the Lord in the day of His wrath. Yet the Revelation has not unfrequently been expounded, as if each vision followed in order of time that which had preceded. Hence hopeless perplexity has been the result.

Another principle of arrangement equally important to be remembered is this : that the final event of blessing (for blessing is the ultimate object of God in all His dealings with His people) is frequently mentioned *first*, before the evil or sorrow is pourtrayed through which that blessing is to be reached. The second chapter of Isaiah affords an example of this. The first part of that chapter describes Israel's Millennial blessing, when "the mountain of the Lord's house shall be established in the top of the mountains, and all nations shall flow unto it, when nation shall not lift up sword against nation, neither shall they learn war any more." After the chapter has thus described the final blessing, it suddenly turns to the day of Israel's iniquity and transgression, and describes the coming of the Day of the Lord which is to *precede* and usher in the Millennial rest. The sixth chapter of the Revelation affords a similar instance. The first verses describe the Millennial triumphs of the Lord going forth conquering and to conquer ; the rest of the chapter describes the preceding judgments that are to usher in the Day of His appearing. The like may be said of all the succeeding visions in the Revelation ; they each begin by first recording some memorial of the scene of final triumph, before they speak of the evil and judgment that are to precede.

It is also necessary to distinguish very carefully between the different periods of man's history as given in the Scripture. They are five in number. The last of these periods is an unchanged condition of eternal blessing; but each of the other four is terminated by a direct interference of God in judgment.

The first of these periods extends from the Creation to the Fall. It was terminated by the exclusion of Man from Paradise.

The second extends from the Fall to the Flood.

The third extends from the Flood to the Second Coming of the Lord Jesus, when God will again interfere as at the Flood, to arrest the progress of human things by an act of judgment.

The fourth is the Millennial period, which will be marked by a general rebellion of nations at the close: met by a final act of God in judgment. Here the history of the Adamic earth ends.

The fifth period is everlasting, and commences by the creation of the New Heavens and New Earth wherein righteousness is to *dwell*.

Due discrimination between these periods is of the last importance in reading the Scripture. If we were to apply to the present condition of human life the principles which were once true of man in Paradise, who would not instantly detect the falsehood? But the error is scarcely less, if we fail to discriminate between the condition of human life in the Millennium and its condition now. In the Millennium, Satan is to be bound—Christ and His truth to be supreme—Israel converted, and made a national witness for God—the nations, minutely regulated by the governmental power of Christ. To confound such a period with one that is carefully and designedly marked in the word of God by characteristics the very opposite to these, is an error scarcely less delusive than to suppose that man is now in Paradise. Yet this mistake has been continually made in the exposition of Scripture. How constantly, for example, are Israel's triumphant Psalms, which they will sing in the Millennium, used as if they were strictly applicable to the condition of humanity now. Men are asked to believe that the world, and all that dwell therein should now rejoice before the Lord, because He has arisen to help all the meek upon earth (Ps. xxxvii. 2), because He is judging the world with righteousness, and the peoples with equity (Ps. xcvi. and xcviii.), because He is opening His hand and satisfying the desire of every living thing (Ps. cxlv. 16), because He hath remembered His mercy and His truth toward the house of Israel, and all the ends of the earth have seen the salvation of God. (Ps. xcviii. 3.) This, according to the present use that is made of

Scripture, is what men are asked to believe. But all is a delusion. Reverse the picture, and you have the truth. Satan is the god of this world, the ruler of the darkness of this present age—meekness is crushed—righteousness oppressed—Israel scattered—"darkness covering the earth, and gross darkness the peoples," and men are obliged to hear and respond to the groan of creation.

Nothing can be more disastrous than the effect produced on unconverted minds, especially if discontented or tending to infidelity, by telling them that they ought to rejoice in the present arrangements of human things. To teach men when smarting under oppression, or misgovernment, or penury, or some unrecognised presence of Satan's power, that all is proceeding rightly under the hand of God, is to mock their misery, and to delude them with a lie. It misrepresents the character of God, and teaches them to mistake the selfishness of man, or the cruelty of Satan, for the graciousness and goodness of God. God has not said that all things are proceeding rightly, or that they are progressing according to His will. On the contrary, He tells us, that "all the foundations of the earth are out of course:" that "pride hath budded," and is bringing forth its fruits, and that iniquity is for a season to occupy the high places of the authority of earth. Let us tell men this: let us tell them that God is displeased with the present arrangements of the Earth; that He is

INTRODUCTORY OBSERVATIONS. 17

grieved at the abounding misery, and will soon interfere "with mighty hand, and with an outstretched arm" to bring in and establish principles which every heart shall recognise as wise, beneficent and good; although it may be that the hearts of those addressed, will still reject the testimony of truth, yet whether they will hear or no, *we* at least shall be guiltless of misrepresenting the character of God. Our words would not be (what they now are) repugnant both to the convictions of men's consciences, and to the experience of their hearts. It may be that their heart will be softened, and an opening afforded for the declaration of the gospel of grace.

Lastly, it is needful to distinguish very carefully between the Millennial period, and the dispensation that succeeds—a dispensation called in Scripture, "the dispensation of the fulness of times," when New Heavens and a New Earth will be created. The Millennial Earth although visited and reigned over by Christ and the risen saints (who will themselves dwell in heavenly places not made with hands) will still be essentially the same earth as that in which we now dwell, inhabited by men in unredeemed bodies liable to sickness and death; men, who, even when sanctified by grace, will still have to say as believers now do, "that in them (that is, in their flesh) no good thing dwelleth." The removal of Satan and his temptations will not change the character of the flesh: "the

C

mind of the flesh, which is not subject to the law of God neither indeed can be," will still be found in all except the glorified: nor will the diffusion of blessings from above, and the repression of decay, destroy the consciousness that the principle of physical corruption is still lurking in every created thing. Death, the last enemy is not to be destroyed until the thousand years are finished. (See 1 Cor. xv. and Rev. xx.)

The reign of Christ over men in this Earth, will indeed make manifest the wonderful resources of His grace and power in remedying sorrow, and subduing evil. It will prove that the natural and providential and spiritual blessings of God, so abused in the dispensations that have preceded, *can* be used, even in this now groaning Earth, for His glory, and for the benefit of those for whose sake they are given. It will be seen how different the condition of human life is, when the plans by which Satan has hindered human happiness, will be supplanted by agencies which God will introduce for blessing, and the wisdom and goodness of Christ will be known in contrast with the evil which, under unregenerate man, has hitherto "destroyed the earth." The Millennium also will be the great harvest-time of the earth, when millions of unconverted souls, will, through the Gospel, be gathered into the garners of God. Nevertheless the Millennium will not be a scene of perfect or everlasting blessing, and nothing that

INTRODUCTORY OBSERVATIONS. 19

is not perfect can satisfy God. It will afford another attestation to the truth of the Lord's own saying, that new wine must not be put into old bottles. The blessings of redemption, are not finally to be connected with anything that is not NEW. Therefore the Millennial Heavens and Millennial Earth pass away, and no place is found for them, and He that sitteth on the Throne will say, "Behold I make all things new," and New Heavens and a New Earth will be created wherein righteousness dwelleth, and into which, the Holy City, the Bride of the Lamb, will descend.

It will be a scene into which Satan will never enter. No apostasy will be there—no repression of evil*—no remedial agency against sorrow—for there will be no grief to be soothed, no sickness

* The Millennial Dispensation is not intended to be one of *perfected* blessing. That which gives it its distinctive character is, that it is a period in which Christ assumes a definite sovereignty (see Dan. vii.) over the Adamic creation *in order that* "*He may put down all enemies:*" and He will retain the power of the Millennial kingdom until He has effected this. "He must reign till he hath put all enemies under his feet. The last enemy that shall be destroyed is death." (See 1 Cor. xv. 25, and Rev. xx. 14.) The Land of Israel—Immanuel's Land—will be "full of the knowledge of the Lord, as the waters cover the seas"; but it must not be supposed that this will be the condition of the whole earth during the Millennium. "The sons of the alien [this is not spoken of Israel, but of some among the Gentiles] shall dissemble unto Me." (See Ps. xviii. 44.) Hence the great Apostasy at the close of the Millennium,

C 2

to be remedied, no sin to be restrained. All will be instinct with life—all made worthy of Him, and

---

after which the Adamic Heavens and earth pass away, and all things are made new.

The Millennial period will be characterised by the restraint and subjugation of Evil, not by its abolishment. Human suffering will be effectually met by remedial agencies; but suffering will not have ceased to be. Hence, in the *symbolic* description of the Heavenly Jerusalem (which will not enter the *Millennial* earth, but, till the new earth is made, will be Επουρανια—*above* the created Heavens) we read that the leaves of the tree were for the "healing of the nations." In the new earth, there will be no "nations" to be healed, nor any sickness.

In reading Revelation xxi., it is most important to remember, that the first eight verses of that chapter should be appended to the twentieth; and that at the ninth verse of the 21st, a new chapter should commence.

The first eight verses of the 21st chapter describe the creation of the *new* earth and the relation of the *New* Jerusalem to *it*. The New Jerusalem descends into the new earth. At the eighth verse the description of the new earth ends; and the succeeding verses recur to a previous period, and treat of the relation of the Heavenly City to the Millennial earth—this description continuing to the end of the 5th verse of chapter xxii.

In Isaiah also, where we find the New Heavens and earth first mentioned, there is a similar arrangement. The 17th verse of chapter lxv. declares the creation of the new heavens and earth. "Behold I create new heavens and a new earth, and the former shall not be remembered nor come into mind." The subsequent verses addressed to forgiven Israel describe their condition in the Millennial earth.

His glory, who is not as the first Adam, earthy; but who is the Second Man, the Lord from Heaven : and if the first Heavens and Earth made in connexion with the first Adam be glorious, what shall be the excellency of that which shall be made in suitability to Him who is the Eternal Son of the Father, God over all, blessed for ever?

CHAPTER II.

NO POETIC EXAGGERATION IN THE LANGUAGE OF SCRIPTURE.

THERE exists in every human heart an innate desire after something more perfect and beautiful than creation in its present condition any where supplies. The eye beholds no form that is quite faultless; it sees no scene in sea or in earth that is complete in beauty. Man, ever longing after something more excellent than the realities of earth afford, thus furnishes an abiding evidence of the ruin wrought by the Fall, and proves that (since Eden was lost), natural perfectness has ceased to be, and man ceased to be satisfied.

In order to remedy this, the powers of taste and imagination are employed. The painter for example, seeing nothing in creation that is quite perfect — finding the beauties of nature scattered, and never existing in complete and harmonious combination—strives, by the power of imagination, to collect and to combine these scattered elements, and thus to produce a whole more perfect than reality any where affords. This has been called the Poetry of painting—its essence being *ideality* or *fiction*, whereby it leads from the real, into an unreal, world. The poet seeks after the same thing. Poetry is not a mere metrical arrangement of words. It is an attempt, by means of language, to present to the mind a picture heightened above reality, and to add, for the sake of adornment, circumstances which are not found in the facts, or in the scene described. A bare narration of facts, would be mere history; and this is so seldom welcome, that even the historian, in order to please, is continually tempted to embellish his narration by imagined sentiments, or heightened colouring. In doing this, he so far ceases to be an historian, and becomes a Poet. The essence therefore of Poetry is ideality or fiction: the secret of its pleasing is, that it leads into an unreal world.*

Of the danger and sin of lingering in an ideal world, it is not now my object to speak. The realities of human life are around us: there is reality in our relation to God here, as well as in the new creation. To tempt from these realities into an unreal world, has ever been one of the most successful efforts of Satan. Men know that they are pleasing themselves with a dream; but it is illusion they seek: they seek it as one of the opiates of life. They are not deceived by poetry;

* "The essence of poetry," says Dr. Blair, is "supposed by Aristotle, Plato, and others, to consist in fiction." Sir Joshua Reynolds also, in his remarks on the poetry of Painting, teaches the same thing.

they know well that it supplies a heightened and unreal picture, and they never think of mistaking it for truth.

But herein consists the evil of saying, that Poetry is found in the word of God. If by this were meant merely, that figurative language, and rhythmical arrangement are found in Scripture, that would be very true. But more than this is intended. It is meant that, as in human compositions, so also in Scripture, the language is heightened, so as to leave on the mind an exaggerated impression—an impression that far exceeds the reality of truth. On this ground the terms of Scripture-description are continually subjected to a process of abatement: much, it is said, must be allowed for embellishment: and thus, each reader rejecting or retaining for himself what it suits him to reject or retain, the Scripture becomes nothing more than a flexible rule, to be bent according to the inclination of those who ought to find in it a rigid and unyielding guide. We may therefore unhesitatingly say, that there is no poetry in Scripture; for there is no description, no statement, which, if received in all the fulness of its meaning, would leave on the mind an impression that would by one jot or tittle exceed the reality. The impression received by us might fall very far short of the truth, but certainly would not exceed it.

The consequences of believing that the exaggerations of Poetry are to be found in Scripture, have been most disastrous. Indeed, it is difficult to see how the great truth of *verbal* inspiration can consistently be maintained in connexion with such a notion: for it seems like mockery to say that each word is carefully inspired by the Holy Ghost in order that the meaning may be conveyed with rigorous precision, and at the same time to speak of poetic embellishment or exaggeration. And the error is the worse, because it necessarily affects those parts of Scripture most, which we ought chiefly to receive with reverent caution and exactness—I mean those parts which speak of the yet future periods of Divine interference in the Earth. These periods, in consequence of being seasons of superhuman and miraculous agency, when the Lord will, in a manner that never yet has been, "make bare His holy arm," are described in language, which must necessarily be of heightened character, in order to convey any just notion of events which exceed every thing which man's experience has hitherto known. Thus the very passages which are of such transcendant importance, because they teach us of that wonderful period when the great results of redemption in relation to this Earth are to be accomplished by the direct interference of almighty power—passages therefore in which of necessity the language bears some proportion to the magnitude of the events described—these are the very parts of Scripture which are not believed, because the force of the language is explained away under pretence of poetic embellishment or exaggeration.*

Let us take as an example the words in which Daniel describes the time when Israel shall be delivered and forgiven; when "there is to be a time of trouble such as never was since there was a nation even to that same time;" * * * when "many of them that sleep in the dust of the earth shall awake, some to everlasting life, and some to shame and everlasting contempt; and they that be wise shall shine as the brightness of the firmament; and they that turn many to righteousness, as the stars

---

* There is something revolting and indeed blasphemous in such a passage as the following. It is from Dr. Blair, in his chapter on the Poetry of the Hebrews: "Of the sacred poets, the most eminent are the author of the Book of Job, David and Isaiah. In the compositions of David there is a great variety of style and manner. In the soft and tender he excels; and there are many lofty passages in his Psalms. But in strength of description he yields to Job, and in sublimity he is inferior to Isaiah. The most sublime of all poets, without exception, is Isaiah. Dr. Lowth compares Isaiah to Homer, Jeremiah to Simonides, and Ezekiel to Æschylus. Among the minor prophets, Hosea, Joel, Micah, Habakkuk, and especially Nahum, are eminent for poetical spirit. * * * The poetry of Job is highly descriptive. It abounds in a peculiar glow of fancy and in metaphor."—*Blair's Essays.* It is obvious that persons who so think and write of Scripture, can only in name receive them as the testimony of the Holy Ghost. If they read them with these thoughts, they may expect to find in them "a gin and a snare" to their own souls.

for ever and ever." This passage is supposed by Porphyry, Grotius, and others, to have been fulfilled when the Maccabees were conflicting with Antiochus and his successors — the language of course being considered hyperbolical.

The manner in which the eighteenth Psalm is commonly interpreted, affords a no less marked instance of perversion. That Psalm is a prophetic description of the period at which the power and majesty of the Divine glory will be put forth, in order to deliver the remnant of Israel, who will then be forgiven and treated as part of the mystical body of Christ: "Then the earth shook and trembled; the foundations also of the hills moved and were shaken, because He was wroth. There went up a smoke out of His nostrils, and fire out of His mouth devoured: coals were kindled by it. He bowed the heavens also, and came down: and darkness was under His feet. And He rode upon a cherub, and did fly, yea, He did fly upon the wings of the wind. * * * * The Lord also thundered in the heavens, and the Highest gave His voice; hailstones and coals of fire. Yea, He sent out His arrows, and scattered them; and He shot out lightnings, and discomfited them. Then the channels of waters were seen, and the foundations of the world were discovered at thy rebuke, O Lord, at the blast of the breath of thy nostrils." These words so solemn, and so incapable of being interpreted of any event excepting one, have nevertheless been continually explained as if accomplished, either at the deliverance of David from his enemies, or at the Resurrection — whereas neither of those events were accompanied by any such manifestation of Divine glory. The Resurrection and Ascension were events accomplished with peculiar secrecy. They were the silent withdrawal of Christ from a world that had rejected Him; and the commencement of a testimony of grace, the like to which never before had been. No arrows were sent forth to scatter—no lightnings to discomfit—no fires to consume.

In the prophecy recorded in the twenty-fourth chapter of Matthew, our Lord, after referring to the words of Daniel which have just been quoted, respecting the unequalled season of tribulation, says: "Immediately after the tribulation of those days shall the sun be darkened, and the moon shall not give her light, and the stars shall fall from Heaven, and the powers of the heavens shall be shaken. And then shall appear the sign of the Son of Man in heaven, and then shall all the tribes of the Land mourn, and they shall see the Son of Man coming in the clouds of heaven with power and great glory. And He shall send His angels with a great sound of a trumpet; and they shall gather together His elect from the four winds, from one end of heaven to the other." These words are still supposed by many to have been fulfilled at the time of the destruction of Jerusalem by the Romans.

The conclusion of the sixth chapter of Revelation reveals in a symbolic vision, the circumstances of the same great day of visitation: " I saw * * * and there was a great earthquake; and the sun became black as sackcloth of hair, and the whole of the moon became as blood ; and the stars of heaven fell unto the earth, even as a fig-tree casteth her untimely figs, when shaken by a mighty wind. And the heaven was separated from its place as a scroll when it rolleth itself together; and every mountain and island were moved out of their places. And the kings of the earth, and the great men, and the chief captains, and the rich men, and mighty men, and every bondman, and freeman hid themselves in the caves and in the rocks of the mountains ; and they say to the mountains and the rocks, 'Fall on us, and hide us from the face of Him that sitteth on the throne, and from the wrath of the Lamb: because the great day of His wrath hath come ; and who is able to stand ?" These words have been commonly referred to the period when Constantine and the Roman Empire assumed the profession of Christianity. Such error seems almost inexcusable —but surely it never would have obtained credence, unless lax thoughts respecting the language of Scripture had first been received. It may, I believe, be safely said that the thoughts that have prevailed respecting the " poetry" of Scripture, have nullified the practical use, or rather I might say, falsified the testimony, of two-thirds of the word of God.

It seems to have been imagined, that whenever the Scripture employs figurative words or symbols instead of simple language, that its statements at once become shadowy and unreal. But even in ordinary human intercourse, men employ sometimes symbols and sometimes figurative language for the communication of plain substantive facts. Suppose an Indian were to stand before us, and breaking his arrows and his bow, were to bury them beneath a tree that was regarded as the emblem of peace—we should understand that his action was symbolic, and intended to signify, that he had ceased from warfare and had resolved on peace: or if he used words and said, "Let us sit down together by the river of peace and drink of the water of quietness," we should understand that his words were figurative, but intended to express his desire after peace: nor would his meaning be conveyed at all more clearly, if, instead of using figurative words or symbolic action, he were to say in simple language. "I desire peace with you." It would be a strange mistake to suppose, that because the action was symbolic, or the words figurative, that therefore no plain intelligible fact was intended. Yet this is the mistake into which many have fallen; and virtually their words imply, that literal facts are never conveyed, except when no figure nor any symbol is employed as the medium. The Scripture employs simple language, figurative language, and symbols, to reveal its truth; but whichever of these be used as the medium, that which is communicated is or will be when accomplished, a REALITY—in no respect falling short of the fulness of the description given. Thus the restoration of Jerusalem and her people to the Divine favour, which will be, when accomplished, a literal fact, is thus taught in Scripture.

It is taught by simple statement, as in Zechariah xiv. 10: "Jerusalem shall be lifted up, and inhabited in her place, from Benjamin's gate unto the place of the first gate, and from the tower of Hananeel unto the king's wine-presses. And men shall dwell in it, and there shall be no more utter destruction; but Jerusalem shall be safely inhabited."

Again, the same event is taught us in Ezekiel xxxvii. by a symbolic action. The Prophet is commanded to take two sticks, to write on one the name of Judah or the two tribes, and on the other the name of Ephraim or the ten tribes, and then, in the sight of his people, to join the two sticks together, so as for them to become one in his hand. This is the symbol. Its explanation is next given. "Thus saith the Lord God, Behold, I will take the children of Israel from among the Gentiles, whither they be gone, and will gather them on every side, and bring them into their own

Land : and I will make them one nation in the Land upon the mountains of Israel; and one King shall be King to them all; and they shall be no more two nations, neither shall they be divided into two kingdoms any more at all.

Thirdly, the same event is foretold in figurative language thus: "Rejoice ye with Jerusalem, and be glad with her, all ye that love her: rejoice for joy with her, all ye that mourn for her: that ye may suck, and be satisfied with the breasts of her consolations; that ye may milk out, and be delighted with the abundance of her glory. For thus saith the Lord, Behold, I will extend peace to her like a river, and the glory of the Gentiles like a flowing stream : then shall ye suck, ye shall be borne upon her sides, and be dandled upon her knees." (Is. lxvi. 10.) It is manifest that the two last-mentioned passages are not less clear than that from Zechariah, which is in simple language. The difficulty therefore which has been supposed to exist in the interpretation of symbols and figurative language, arises far more from the indisposition of our minds to receive the truth, than from any obscurity in the medium of instruction.

The two Books in which symbols are chiefly employed, are Daniel and the Revelation. In other parts of Scripture, the use of symbols is comparatively rare. The thirty-seventh chapter of Ezekiel already referred to, contains two of the most remarkable instances; there are some in the commencing chapters of Zechariah, and one in the Acts, when Peter saw the sheet let down from heaven. No difficulty *can* exist in the interpretation of the greater part of the symbols of Scripture, because the Scripture itself supplies the interpretation. Thus the two symbols in Ezekiel xxxvii. are interpreted in the same chapter, for we are told that the two sticks on which the names of Israel and Judah were written, represent Israel nationally dead. In the Revelation, the greater part of the symbols are interpreted either in the Book itself, or by connexion with some other part of Scripture such as Daniel. The golden Candlesticks are explained to represent Churches; and the three unclean spirits like frogs that came out of the mouth of the Dragon and of the Beast, and of the false Prophet (see Rev. xvi. 13) are explained to be spirits of devils working miracles that go forth to deceive," etc. In other cases, where explanation is not given, the clearness of the symbol renders the interpretation so self-obvious, that no explanation is required. Thus an angel was seen in the vision "coming down from heaven, having the key of the abyss and a great chain in his hand. And he laid hold on the Dragon, the old serpent, which is the Devil, and Satan, and bound him a thousand years, and cast him into the abyss, and shut him up, and set a seal upon him, that he should deceive the nations no more, till the thousand years should be completed." In this case no explanation need be given. The vision explains itself. The principal and almost only rule to be remembered in the interpretation of symbolic visions is, that the symbol is to be distinguished from the thing symbolised. The first is a sign merely, a sight seen in the vision, an unreal thing; but the thing symbolised is a reality. The golden Candlesticks were signs merely, but the Churches whose honoured position in the earth was thereby symbolised, were realities. The outer court of the Temple seen in the eleventh chapter was a symbol merely: it was not a reality: but that which is symbolised thereby is a reality, viz., Jerusalem—the Holy City, for so the symbol is explained in the passage itself. It would be an equal error therefore, to assign reality to the symbol, or non-reality to the thing symbolised.

As respects the interpretation of the figurative *language* of Scripture, I would again say that even when the Scripture uses the strongest and most vivid figures, the impression left on the mind thereby can never exceed the reality. Moreover no language is to be regarded as figurative, that *can*, without involving an absurdity or impossibility, be interpreted simply. When Israel is spoken of as "borne upon the sides, and dandled on the knees of Jerusalem," we instantly see that the language is figurative, because a literal interpretation would involve absurdity. But it is otherwise when we read, "that the wolf shall dwell with the lamb, and the leopard shall lie down with the kid; and the calf and the young lion and the fatling together; and a little child shall lead them. And the cow and the bear shall feed; their young ones shall lie down together: and the lion shall eat straw like the ox. And the sucking child shall play on the hole of the asp, and the weaned child shall put his hand on the cockatrice' den. They shall not hurt nor destroy in all my holy mountain: for the earth [or Land] shall be full of the knowledge of the Lord, as the waters cover the sea." In this case there is no need to explain the language as figurative, for there is no impossibility in animals again becoming what they once were when originally created. We are told in Genesis that in Paradise they all ate the herb of the field, and until sin entered, through man, death and destruction was unknown amongst them. The promises therefore of this passage we expect to be literally fulfilled when the Lord Jesus shall return and assume His Millennial power. We say that the language is *not* figurative. Nevertheless, the facts described thereby may be symbolic facts, indicating that throughout a wider and more important sphere, there shall be a calming throughout the earth of those evil passions which had, in former Dispensations, filled the world with bloodshed, desolation, and woe.

The rule that language is never to be regarded

as figurative, except when necessity requires it, is founded on an examination of the mode in which *past* prophecy has been fulfilled: for we cannot but conclude, that the mode which God has been pleased to adopt in the fulfilment of prophecies that are past, will be carefully adhered to in the accomplishment of those that are to come.

The most important of the prophecies already fulfilled concern the appearance of the Lord Jesus in humiliation. If we had lived before those prophecies had been accomplished we should probably have regarded much of the language as figurative. We should doubtless have said, that the prophecies respecting the vinegar and the gall—the dividing the garments but the casting lots on the vesture —the smiting on the cheek—the spitting—the riding on the ass and the like, were to be understood as general figurative descriptions of suffering and meekness, but that the words were not to be interpreted with too minute exactness. But they have all been fulfilled to the very letter. The following are the circumstances of the crucifixion.

PROPHECY.	FULFILMENT.
" It was not *he* that hated me that did magnify himself against me, then *I would have hid myself from him;* but it was thou, a man, my equal, *my guide*, and mine acquaintance." Ps. lv. 12.	" And Judas also which betrayed Him, *knew the place*, for Jesus ofttimes resorted thither with His disciples." John xviii. 2.

NO POETIC EXAGGERATION, ETC. 37

"And I said unto them, If ye think good, give me my price; so they *weighed* for my price thirty pieces of silver." Zech. xi. 12.

"And the Lord said unto me, Cast it unto the potter: a goodly price that I was prized at of them. And I took the thirty pieces of silver, and cast them to the potter in the house of the Lord." Zech. xi. 13.

"They shall smite the judge of Israel with a rod upon His cheek." Mic. v. 1.

"They gave me also gall for my meat; and in my thirst they gave me vinegar to drink." Ps. lxix. 21.

"My God! my God! why hast thou forsaken me?" Ps. xxii. 1.

"I hid not my face from shame and spitting." Is. l. 6.

"They pierced my hands and my feet." Ps. xxii.

"What will ye give me, and I will betray Him unto you? And they *weighed* to him thirty pieces of silver." Matt. xxvi. 15.

"Then Judas brought again the thirty pieces of silver, and cast them down *in the temple*, and the chief priests took them, and bought with them the potter's field." Matt. xxvii.

"They took the reed and smote Him on the head." Matt. xxvii. 30.

"They gave Him vinegar to drink, mingled with gall; and when He had tasted thereof, He would not drink." Matt. xxvii. 34.

"Jesus cried with a loud voice, saying, My God! my God! why hast thou forsaken me?" Matt. xxvii. 46.

"Then they spit on Him." Matt. xxvii. 30.

"They crucified Him." John xix. 18.

"He keepeth all His bones, not one of them is broken." Ps. xxxiv. 20.	"These things were done that the Scripture might be fulfilled: A bone of Him shall not be broken." John xix. 36.
"He was numbered with the transgressors." Is. liii. 12.	"They crucified Him, and the malefactors, one on the right hand, and the other on the left." Luke xxiii. 33.
"They part my garments amongst them." Ps. xxii.	"The soldiers took the garments of Jesus, and made four parts; to every soldier a part." John xix. 23.
"They cast lots for my vesture." Ps. xxii.	"But his coat was without seam, woven from the top throughout; they said therefore among themselves, Let not us rend it, but cast lots for it, whose it shall be." John xix. 23.

Can there be more exact fulfilment, than these prophecies have received? Let past fulfilment then supply us with the rule for that which is to come. There is much yet to come, for the Prophets have not merely testified of the sufferings of Christ — they speak also of the glory that is to follow.

I have already spoken of the manner in which

men instinctively turn from the realities around them, and seek relief in an ideal world. The Believer indeed does not in the same way need relief, because he has joys even in suffering for the Truth's sake which gild many a dark scene in life. "Thou hast put gladness in my heart more than in the time when their wine and their oil increased." There is a present knowledge of God—of His kindness, of His grace, and of His truth which comforts in many a sorrow, and enlightens many a path that would otherwise be cheerless and forlorn. Nevertheless, even the Believer needs, and God designs, that he should have the sustainment of anticipative hope. There are things not seen as yet—things waited for, which, whilst they withdraw the heart from the sad realities around, do nevertheless lead into no *unreal* scene. They lead on into the future; but it is a future pourtrayed by the words of the God of truth—words none of which shall fail; "none shall want their mate"—for every word that He has spoken is exact; every promise faithful. It is thus through patience and comfort of the Scripture we have hope.

The Scriptures reveal the happy prospects of the Earth, and of those who are to reign over it with Christ in the age to come. They tell us of the time when converted Israel, and the nations, and creation freed from the bondage of corruption, will together rejoice under the hand of God. They reveal, also, the higher glories of those who shall dwell in heavenly places not made with hands, and who shall share in the glories of Christ risen, and have power to follow Him whithersoever He goeth, whether in Earth or Heaven. (See Rev. xiv.) The Revelation especially presents, in a variety of aspects, the various glories of the risen saints. Thus the prospects of the future, and even the circumstances of the kingdom of glory, are so presented as to give to our anticipations such a definiteness and precision as is almost essential to their being influential anticipations. But if the heightened language in which the Scripture describes these future scenes of blessing is explained away under the pretence of poetic exaggeration, not only is the object of Scripture in giving those descriptions nullified, but an occasion is afforded for their perversion. Accordingly the principles of interpretation which have in modern times prevailed, have not only caused the descriptions of the earth's Millennial blessing to be applied to this present period of creation's sorrow, but have even so perverted visions of future glory as to cause them to be applied to past or present periods of man's worst iniquity. Thus the corruptions of the early post-apostolic Church, the sinful elevation of Christianity in the days of Constantine, and the false expectations of Protestant Christianity in England — a country which is governmentally tampering with, and fostering, almost every form of influential religiousness that is to be found in earth—have been said to be represented by some of the most holy and heavenly visions which the Revelation contains.*

Is it any wonder that under such circumstances confusion and darkness should reign, when we are taught to believe that the Scripture celebrates the period of creation's groan in psalms of thanksgiving, and represents dark periods of man's evil by visions of glory. If such interpretations prevail, there is reason to fear that the heart will soon cease to feel, and the conscience to judge; and the book of Prophecy will be not understood, because the eye of the soul will have become dim. "If the light that is in thee be darkness, how great is that darkness!"

* Thus in the Revelation, the "woman clothed with the Sun," &c., has been said to represent the political elevation of Christianity in the days of Constantine; and those standing on the sea of glass, having the harps of God, and singing the song of Moses, and of the Lamb, have been said to represent the condition of Protestant Christians in England, rejoicing after the conclusion of the Revolutionary wars.—*See works of Mr. Elliott and Dr. Cumming.*

CHAPTER III.

SOME OBJECTIONS TO THE DOCTRINE OF THE MILLENNIAL REIGN CONSIDERED.

BEFORE we proceed to the definite consideration of some of the prophetic parts of Scripture, it may be desirable to consider briefly, some of the objections which have been urged against the Millennial reign of Christ.

It has been said, that if we teach the personal appearing of the Lord, we must suppose that Christ and the glorified saints, instead of having Heaven for their home, and heavenly glory for their portion, will be necessitated to return to earth, and be restricted to earth as their dwelling-place.

It cannot be denied that many who have written on the Millennium, from the earliest times to the present, are justly chargeable with error on this subject. Very soon after the Apostles died, the glory of the risen saints in "heavenly places," was confounded with that of Israel, and the nations on earth. No adequate distinction was drawn between the Jerusalem which is above, and the Jerusalem which Israel is to inhabit on the earth. Thus many were stumbled, and Jerome, in the fourth century, denounced the doctrine, and spoke "of the fable of the thousand years' reign." It had indeed become a fable as taught by men, but the fancies of men are not to be confounded with what the Scriptures reveal.

The Scriptures teach, that Heaven will always be the home of Christ and of the risen saints. He is gone to prepare mansions for them in His Father's house; and into those mansions they will be received, and will have them for their own for ever. Hence they are prospectively called in the Scripture, "saints of the high places"* (Dan. vii. 18)—"stars" (Dan. viii. 10)—"Host of the Heavens" (Dan. viii. 10): but the same chapters which thus describe their heavenly glory, speak of their reigning with Him to whom dominion "over all peoples, nations, and languages" is to be given. (Dan. vii. 14.) They will indeed superintend, together with Christ, the government of earth, but they will not on that account vacate the sphere of heavenly glory above. Angels even now, during this dark dispensation of sorrow daily visit this earth, ministering to the heirs of salvation. Moses and Elijah have already been seen in glory on this earth. Their feet really stood upon an earthly mountain: and their form was visible to mortal eyes; yet they did not dwell

* קַדִּישֵׁי עֶלְיוֹנִין, an expression used four times in Dan. vii. See verses 18, 22, 25, and 27.

on earth; Heaven was their dwelling-place, and they were seen to return into clouds of heavenly glory. So will it be with Christ and the risen saints. They will visit the earth, they will be *seen* on earth; the seat or Throne of Christ's earthly government will be established on earth (for "the Lord of hosts shall reign in Mount Zion, and in Jerusalem, and before His ancients gloriously": see Isaiah xxiv. 23), but neither He nor they will dwell on earth. His glory will be set "above the Heaven." (Ps. viii.) The ladder which Jacob saw did not indicate that Heaven was changed into earth, or earth into Heaven; it was the symbol of connection between two places united in blessing, whilst remaining locally apart. "Ye shall see," said our Lord, "Heaven open, and the angels of God ascending and descending upon the Son of Man." The Millennial earth, therefore, will not be the *home* either of Christ or of His risen saints. When God descended on Sinai, and when His glory was seen, and when He spake to Moses, and legislated for Israel there, His *home* was still in Heaven. So will it be when Christ shall legislate for Israel and the earth from Zion.

Another difficulty has been felt by some in understanding how it can be said of the Lord Jesus, that He has already "all power in heaven and in earth," and yet, at a time still future, be brought before the Ancient of days to be invested with further sovereignty.

This difficulty is easily answered when we remember that He who essentially possesses all power, may for a season *delegate* to others the administration of a certain portion of that power. To the Throne of God all power essentially and eternally belongs. It is the source of every thing, upholds every thing, and ultimately controls every thing. But it may delegate to others the administration of a portion of its governmental rule; and it has done this. When God punished Jerusalem by bringing against it the King of Babylon, it was said to Nebuchadnezzar, "the God of Heaven hath given thee a kingdom, power, and strength, and glory," &c.—and it was further said, that the Empires that were to follow him, should receive the like endowment, and bear rule in all the earth. It was under this delegated power, when vested in Rome, that the Lord Jesus when on earth lived and suffered. When He ascended, it might well be said that "all power was given unto Him in Heaven and in earth," for He entered upon the exercise of all the power of the Throne of the Most High God. But the power which that Throne had delegated, continued delegated still; and so will continue, until the appointed hour comes for its reassumption by the Throne to which it inherently belongs; and then, it will be committed unto Christ, who will be pleased to undertake the definite legislation and government of Israel and of the nations. It will not be a possession of new power, but delegated government recovered and assumed. Therefore, when great voices shall be heard in Heaven, saying, "The sovereignty of the world (ἡ βασιλεία τοῦ κόσμου) hath become the sovereignty of our Lord and of His Christ" (see Rev. xi. 15), it is immediately afterwards said, "We give thee thanks, O Lord God the Almighty, who art, and who wast, because thou hast taken to thee thy great power and hast reigned." This verse evidently refers to delegated power *resumed*.

It has also been asked by some, whether we admit that Christ is now a King. He was ever a King; He was born a King; legislated as a King; entered Jerusalem as a King; was crucified as a King. As the rejected King of Israel He now sits on the Throne of the Majesty in the Heavens, exercises all the power of that Throne, governs Heaven, controls earth, and legislates for His Church. As King also, He will return; destroy the governmental systems of the nations, and undertake the minute governmental regulation of Israel and the earth, without resigning the sphere of His glorious majesty above.

Again it has been objected, that Millennial writers have represented the Christianity of the Millennium as something different from Christianity as now received, and thus teach another Gospel. It is indeed true that not a few, by their unguarded and erroneous statements, have laid themselves open to the charge. But the Scriptures teach no such error. They teach that Israel in the Millennium, will be brought under the *same* "New Covenant" of grace under which we who believe in Jesus now are. They will have the same one finished sacrifice, the same priest passed into the Heavens, the same forgiveness through the same precious blood, the same Spirit, the same union with a risen Lord, the same prospects in "new Heavens and a new earth"; where all the redeemed will meet in the same glory; for they who have Christ, have "all things." There are not two gospels, or two ways, or two ends of salvation. Indeed the character of our dispensation is briefly this. It is one in which all the spiritual and eternal blessings which, by and by, will rest on Israel, are received beforehand by the suffering family of faith. When Israel shall nationally receive the Gospel, the change in external circumstances, will indeed be great. Satan will be bound, creation freed from its groan, and the national government of the earth; be vested in the same hand to which God has already committed the government of His Church. The outward condition of human life will be in harmony then with the blessings inwardly received by faith: "Sorrow and sighing shall flee away" (Isaiah xxxv. 10), and Truth will be triumphant, instead of being rejected and despised. But this change of circumstances will not alter the essentialities of truth. The eternal veri-tics of the Gospel must always be the same—even as gems have the same intrinsic preciousness, whether trodden in the dust, or set in acknowledged excellency in the diadem of the King. Those who will rise in the "*first* resurrection," at the commencement of the Millennium, are called "the Church of the first-born-ones," (ἐκκλησία πρωτοτόκων, Heb. xiii. 23), but they do not constitute the *whole* Church; else they would not be called "the Church of the first-born-ones." The whole Church will not be complete until all those who shall be born into the family of faith during the Millennium shall have been raised and glorified, and joined their brethren in the New Heaven and New Earth where righteousness shall abidingly dwell.

The resemblance, as regards spiritual blessings, between the condition of Israel in the Millennium, and that of Believers now, enables us to remove another difficulty which has been felt by many. It has long been a favourite habit with many Believers, to interpret of themselves many passages in Isaiah and the Old Testament Prophets which speak of the blessings that will attach to the people of God in the Millennial age; and when told, without explanation, that they are wrong in their interpretation, and that these passages belong not to them, but to Israel in another Dispensation, they have been justly offended, and felt as if they were being defrauded of some of their most valued sources of consolation.

It is indeed true that it is wrong to *interpret* of the people of God now, passages that describe the outward physical blessings which will be granted in the Millennial age. *Interpretation* is to be distinguished from *application*. We can only *interpret*, when every jot and tittle of the description is applicable to the persons of whom it is spoken ; and that will never be the case, except when we use the passage in its entirety of those for whom it is *primarily* intended by the Holy Ghost. They who are in circumstances of sorrow and reproach, cannot *interpret* of themselves, passages which describe a condition of outward prosperity and triumph; but if the *spiritual* blessings are essentially the same in the case of those who are suffering and in the case of those who are triumphant, there must be parts in the descriptions of those who triumph, which may well be borrowed and *applied* to those who are in the sorrow. We cannot in this present dispensation say that there is "no lion in the path of the redeemed" (Is. xxxv. 9), or that "sorrow and sighing have fled away" (Is. xxxv. 10), or that "tears have been wiped away from all faces" (Is. xxv. 8), or that "the Lord of Hosts is reigning in Mount Zion and in Jerusalem and before His ancients gloriously" (Is. xxiv. 23), but we can say "the desire of our soul is to Thy name and to the remembrance of Thee"; and again, "Thou wilt keep him in perfect peace, whose mind is stayed on Thee: because he trusteth in Thee.

Trust ye in the Lord for ever: for in the Lord Jehovah is everlasting strength." Multitudes of such statements we may *apply* to our own condition now, and from Israel's future songs of thanksgiving, borrow the language of our present confidence.* All the *spiritual* blessings of the Millennium we forestall.

It is therefore of the last importance in reading the Scripture, that we should distinguish the different periods of which it treats; otherwise it is impossible to fulfil the commandment of the Apostle, "rightly to divide the word of truth."

---

* Every *spiritual* blessing spoken of in Millennial passages may be taken by ourselves, and we may also use every passage which speaks of the abstract character of God. Thus when it is said: "A father of the fatherless and a judge of the widows, is God in His holy habitation," this will be *manifestly* true in the Millennium — the fatherless will be befriended, and the widow cease to be oppressed. But it is true even *now* that God is this, though He may for a season withhold the interference of His hand; and faith can recognise this and rejoice, even whilst it suffers. Thus the love and sympathy and power of Christ were as real during the time that He refrained from going to the sorrowing sisters of Lazarus, as afterwards when He went and put forth His power, and relieved them of their anguish.

It must also be remembered that we do not deny the truth of Christ's *spiritual* reign, because we say that He will also undertake the outward regulation of the nations. He will reign spiritually in the hearts of His people in the Millennium, even as He now does in all those who are His.

The periods of which the Scripture treats are, as I have already said, five in number.

I. The period of man's innocency in Paradise.

II. The period from Paradise to the Flood.

III. From the Flood to the Second Advent of the Lord.

IV. The Millennial period.

V. The New Heavens and New Earth.

If these periods are confounded—if, for example, we *interpret* of the present period those parts of Scripture which belong to the Millennial age, nothing but confusion and error can ensue. We might just as well assert that we were living in Eden, surrounded with the paradisiacal blessings of Adam in his innocency, as say that we are holding now, or that we are intended to hold now, that position of strength and supremacy in the earth which is reserved for Israel in "the age to come." Before the Apostles died, the temptation to the Church was, "to reign as Kings before the time" (1 Cor. iv. 8), and, as soon as the Apostles *had* died, they yielded to the temptation, and renounced the garb of Nazareth for the goodly garments of Kings' Courts. The Lord Jesus was baptized at Bethany;* when near Jerusalem He may be said to have lived at Bethany; was anointed for His burial at Bethany, and from Bethany ascends into Heaven. Bethany means the house of the poor or afflicted One. It denoted a position that Christ loved: but His people have hated it. They have resolved to exalt themselves, and to rule. It is no wonder therefore that they should have sought to bend Millennial Scripture to their purposes of self-exaltation, and have laboured to sanction their false position, by perverted applications of the word of God. They have wished that "the glory of Lebanon" should come unto them, "the fir-tree, the pine-tree and the box together"; they have wished to suck the milk of the Gentiles and to suck the breast of kings, and so they have tried to thrust themselves into the place of Jerusalem in her future excellency, and have sought to gather around themselves the riches and glory of the world, *before* the time comes for its being hallowed to the Lord of Hosts; *before* the officers of earth are peace, or her exactors righteousness; *before* the sovereignty of the earth has become the sovereignty of the Lord—*before* He has taken to Him His great power and reigned. The sin of this has, for the most part, been unrecognised even by Christ's

* Such is unquestionably the right reading in John i. 28. There were therefore two places named Bethany; one near the Jordan; the other near Jerusalem, on the Mount of Olives.

true people. But until it is recognised, confessed, and relinquished, we cannot expect recovery from our present state of scattering and confusion.

In rightly dividing the Scripture, it is also of essential importance to distinguish certain points of subdivision, in the history of the period in which we now are.

The first, is the call of Abraham, and the consequent separation of the nation of Israel.

The second, is the era of Nebuchadnezzar, when certain successive Gentile Empires were prophetically appointed to hold supreme power in the earth, during the whole period of Jerusalem's abasement.

The third, is the formation of the Christian Church at Pentecost.

As soon as the waters of the Flood were withdrawn, the evil of man again began to manifest itself throughout the earth; and the progress of Idolatry was so rapid, that even when Abraham was born, his family were serving other gods. Universal idolatry would have reigned, unless God had interfered in calling Abraham. The rest of the world went onward in their course of darkness, but the family of Israel were separated and brought into the knowledge of the true God. From that time, the history of the family of Abraham forms a distinct subject in the record of Scripture.

After many years of trial, Jerusalem was proved unworthy of that place of supremacy in the Earth which she had held for a short period during the reign of Solomon. She was accordingly removed from that position, and the place of supreme authority was assigned to Babylon; and after Babylon, to three other Empires which successively followed. The history of these kingdoms is chiefly given in the Book of Daniel. The whole period of their dominance is called in Scripture the "Times of the Gentiles," and is coincident with the period of Jerusalem's abasement. "Jerusalem," said the Lord Jesus, "shall be trodden down of the Gentiles till THE TIMES OF THE GENTILES be fulfilled." The history of this period as given in the Scripture is divided thus:—

I. From Nebuchadnezzar to the destruction of Jerusalem by the Romans.

II. The period that intervenes between the destruction of Jerusalem and the return of the Jews in unbelief to their own Land.

III. The period that intervenes between the return of Israel in unbelief and the commencement of the Millennial reign.

Of these periods the second is almost passed over in silence in the historic prophecies of Scripture; for the Scripture gives no *detailed* history of the Gentile nations except in connection with Jerusalem. It mentions no personages, nor localities, and supplies no dates during the time of the national extinction of Jerusalem. But the third, or future period, is treated of with peculiar precision.

Indeed, all the most important parts of the visions of Daniel and the Revelation concern this *future* period.

Thirdly, we have to consider the history of Christendom, or the Professing Church, which commenced at Pentecost. Its history from that time proceeds concurrently with that of Israel and the Gentile nations. I scarcely need say that it is of essential importance to distinguish between those parts of Scripture which severally belong to the Jews, the Gentile Empires, and the Professing Church. The history of these bodies is kept carefully separate in the Scripture, and the habit of confusing them, which has almost universally prevailed among prophetic writers, has been one great cause of the difficulties which have perplexed their systems. We can easily conceive how great the confusion must be if we assign to the Professing Church prophecies which belong either to the Jews, or to the ruling Gentile nations. Yet this has continually been done.

From the moment when the Jews were first made a separate people on to the time when they shall look on Him whom they have pierced, as well as afterwards, during the Millennium, they have in the Scripture a distinct history of their own. It is by interpreting, therefore, of Judah and Jerusalem prophecies which are avowedly written of Judah and Jerusalem that one great source of perplexity is avoided.

The prophecies which refer to Jerusalem after the Jews re-gather there in unbelief (some of the most simple as well as most important which the Scripture contain) have been peculiarly neglected by prophetic writers. It is for this reason that I have selected for the first subject of our consideration the concluding chapters of Zechariah. Whatever difference of judgment may exist as to the exposition of particular expressions or verses, I can scarcely conceive that any one can candidly read these chapters without being constrained to allow:

I. That the Jews will be as a nation converted.

II. That they will be, at the time of their conversion, in their own Land and City, and consequently must have returned in unbelief.

III. That they are again punished, after their return, by Gentile nations being once more gathered in siege against their City.

IV. That the Gentile nations so gathered are *there* (that is to say, in the Land of Israel) to be destroyed.

V. That they will be destroyed by the personal intervention and manifestation of the Lord in glory.

These are events of no trifling moment. If they be true, if they can be definitely learned from these simple chapters, and be established as ascertained truths, we shall have gained no unimportant light. These facts will be to us as landmarks. They will steadily guide us in our subsequent enquiries, and we may safely say that nothing can be true that contradicts them—every system must be valueless that makes no room for them.*

* It is worthy of remark, that writers whose views have for the most part little harmonised with those we are now advocating, have been constrained to admit the futurity of that part of Zechariah which we are about to consider. Thus Scott, in his Commentary, quotes with approbation, the following passage from Lowth: "The former part of this chapter * * * * relates to an invasion made upon the inhabitants of Judæa and Jerusalem in the latter times of the world; probably after their return to, and settlement in their own land."—*Lowth, quoted by Scott.*

CHAPTER IV.

ON ZECHARIAH XII. AND XIII.

THE twelfth chapter of Zechariah is one of peculiar simplicity. It treats, indeed, of the *future*, but its statements are so plain that instruction could not be given more simply by the most direct historic narrative of the past. It commences by the Jehovah of Israel declaring His title to almighty and creative power. At the period of which this chapter treats that title will have been denied. One will have arisen in the midst of Israel of whom it is written that "he will do according to his will, and he shall exalt himself and magnify himself above every god, and shall speak marvellous things against the God of gods, and shall prosper till the indignation be accomplished; for that that is determined shall be done."* (Daniel xi. 36.) Multitudes, both in Israel and among the nations, will

* It is important to notice the expression, "till the indignation be accomplished," *i.e.*, God's indignation *against Jerusalem*. It is a period frequently referred to in Daniel, called sometimes, "end of the indignation," as in chap. viii. 19. The occurrence of these words identifies these passages in Daniel, *as to time*, with the chapter in Zechariah we are considering, for that also treats of the last end of the indignation against Jerusalem.

have followed him, owned him, and concurred in saying both of Jehovah and of His Anointed, "Let us break their bands asunder, and cast away their cords from us." It will be a time of abounding blasphemy, and therefore many of those parts of Scripture that pertain to this period peculiarly bear testimony to the governmental and creative power of God. "The burthen of the word of Jehovah for Israel, saith Jehovah, which stretcheth forth the heavens and layeth the foundations of the earth, and formeth the spirit of man within him."

This, His power, which He has so long used against Israel because of their sins, He will now be about to put forth on their behalf. *That* is the special subject of this chapter. It speaks of Jerusalem as surrounded by unnumbered hosts of nations, who, wishing to blot out the name of God from the earth, and hating Jerusalem because of its association with that name, will confederate against Israel, and say, "Come and let us cut them off from being a nation, that the name of Israel may be no more in remembrance." But the appointed hour will have come for the God of Israel to interfere. "Behold, I will make Jerusalem a cup of trembling unto all the peoples round about." "I will make Jerusalem a burthensome stone for all the peoples ;* all that burthen

---

* "Peoples," not people, is the right translation throughout these chapters. "Peoples" in the plural always appropriates the expression to Gentiles.

themselves with it shall be cut in pieces, though all the peoples of the earth be gathered together against it."

These words sufficiently indicate the mighty strength and multitude of these nations that will then be congregated against that apparently doomed city. The gathering of these hosts is not unfrequently referred to in the Scripture, and always in language calculated to impress the mind with the peculiar magnitude of the power to be displayed in this last great effort of man under Satan. In the Revelation, for example (ch. xvi. 14), it is said that "spirits of devils working miracles shall go forth to gather the kings of the whole world* to the battle of that great day of God Almighty." Joel also speaks of the same mighty confederation: "Proclaim ye this among the Gentiles, prepare war, wake up the mighty men, let all the men of war draw near, let them come up; beat your ploughshares into swords and your pruning-hooks into spears; let the weak say, I am strong. Assemble yourselves, and come, all ye Gentiles, and gather yourselves together round about." (Joel iii. 9—12.) And what will Jeru-

---

* That is, the Roman world, [ἡ οικουμενη] *orbis terrarum.* Compare Luke ii. 1. The words, "of the earth," are an interpolation, and should be omitted. See Tregelles' "Version of the Revelation," printed by Bagster, a book that should be possessed by all who desire to read the Revelation carefully.

salem appear in their sight? It will be even as
nothing. The tribe of Judah, too, will be with
these nations fighting *against* Jerusalem,* so that
it will indeed be to that chastened city a day of
weakness and of bringing low.

But Jerusalem is not to be forsaken for ever. It
is the place which Jehovah hath chosen to set
His name there. He has even said that His eyes
and His heart are there perpetually. The nations
may be allowed to trample on it for a season; but

---

* That is, that part of Judah which will be dwelling in
Judæa, without the walls of Jerusalem. The literal translation of the second verse is as follows. It is given almost
correctly in the margin of our Bibles. "Behold, I will
make Jerusalem a cup of trembling unto all the peoples
round about, and also against Judah shall it (or He) be
in the siege against Jerusalem." Whether we take "it" as
referring to the cup of trembling or "He" as referring to
Jehovah, the general sense will be the same. It is obvious,
not only from this verse, but from all the rest of the chapter,
that Judah is regarded as in the camp of the enemy. They
will not, like the inhabitants of Jerusalem, be defended by
walls, and therefore terror, probably, will cause them to unite
with the Gentile hosts. The dreadfulness of these hosts is
remarkably described in Joel ii. "Blow ye the trumpet in
Zion, and sound an alarm in my holy mountain: let all the
inhabitants of the land tremble, for the day of the Lord
cometh, for it is nigh at hand. A day of darkness and of
gloominess, a day of clouds and of thick darkness, as the
morning spread upon the mountains; a great people and a
strong, there hath not been ever the like, neither shall be
any more after it, even to the years of many generations."

they are strange nations to Him, ever symbolised in Scripture by wild and evil monsters (θηρια)— nations uncalled by His name, who, just at that very moment, will have said, "Let us break their bands asunder, and cast away their cords from us." *We* have especial need to remember this, for they are the very nations among whom we dwell. All the nations of the Roman earth, from Great Britain to the Euphrates, and beyond, (for the gathering is said to be from the whole Roman world πασα ἡ οικουμενη,) will have sent the flower of their strength to Armageddon. The horsemen of western Europe, and of Syria and Arabia in numbered squadrons will be there, and against them, (for these apparently are the pride of their glory) against them first, Jehovah will direct His hand. "In that day I will smite every horse with astonishment, and his rider with madness, and I will open mine eyes upon the house of Judah, and will smite every horse of the peoples with blindness." It will be literally true. Astonishment and blindness will fall on every horse—madness on every rider. This will commence the ruin of those doomed hosts. But with Judah it will be otherwise. Judah will indeed be in the midst of these doomed nations, and like them will appear in siege against Jerusalem. But at the very moment when He thus smites these hosts of the alien, He "opens His eyes upon the house of Judah," and their hearts are touched with repentance. The sight of the maddened squadrons around them, combined with the sense of their own marvellous exemption from the same well-deserved infliction, will probably be the means whereby their hearts are softened into repentance: but these means would have been in vain, their hearts would have been proof against every mercy, and against every terror, if His eyes and His power had not been directed towards them in grace and pardoning love. "He will open His eyes upon the house of Judah," and their hearts will be softened.

The first evidence of the repentance of Judah is, that they recognise Jerusalem to be indeed the City of God. "The Governors of Judah shall say in their heart, The inhabitants of Jerusalem shall be my strength in the Lord of Hosts their God." And as their hearts utter these words—as soon as they seek their strength in those with whom God is, then they are strengthened. "In that day I will make the governors of Judah like a hearth of fire among the wood, and like a torch of fire in a sheaf; and they shall devour all the peoples round about, on the right hand and on the left: and Jerusalem shall be inhabited again in her own place, even in Jerusalem." Judah shall be as the torch, and the nations as the sheaf, and they shall be consumed as stubble before the fire. It will be as the day of Midian again, and Jerusalem shall be delivered.

It is grace, sovereign unmerited grace, that will

work this deliverance; and therefore Judah, even they who had been in confederation with the foe, will be delivered *first.* "The Lord also shall save the tents of Judah FIRST, that the glory of the house of David, and the glory of the inhabitants of Jerusalem do not magnify themselves against Judah." He knows the tendency of the human heart to exalt itself, even at the expense of others, and will therefore provide that no such manifestation of evil should mar the excellency of that day of mercy. It will be true that Judah will be found in a place more rebellious and more evil than that of Jerusalem; lest therefore Jerusalem and her people should seek to despise or taunt their brethren for their defection, grace, which reaches even to the uttermost, saves the most evil *first*, and thus Judah is strengthened even before Jerusalem is saved, that so every heart might be comforted, every thought of pride silenced.

But Jerusalem and her people will in their turn be strengthened also, and that with no ordinary strength. They shall be so strengthened, that "he that is feeble among them shall be as David, and the house of David shall be as God, as the angel of the Lord before them." The house of David shall be as God, for the Son of David, even the Lord Jesus in His glory, shall then assume the headship of David's house. He will appear in His glory as God. "His feet shall stand in that day upon the Mount of Olives," His saints surrounding

Him. "The Lord my God shall come, and all the saints with Thee." (Zech. xiv. 5.) He will come "to render His anger with fury, and His rebuke with flames of fire." No marvel therefore that *they* should be strengthened with whom He connects Himself in the power of His salvation, or that *they* should be destroyed, who are gathered against Jerusalem and its King. "There shall come out of Zion the Deliverer, and turn away unrighteousness from Jacob."

It will be an hour of triumph to Israel, such as they never before have known—greater than when they quitted Egypt—greater than when they entered the land, and the walls of Jericho fell down. But great as their triumph will be, great as will have been their individual might, (for "he that is feeble among them will be as David,") yet when they return from their victory, this, their glorious day of triumph, will end in self-abasement and in tears. On former occasions, when "Jeshurun had been made to ride on the high places of the earth, he had waxed fat and kicked; then he forsook God which made him, and lightly esteemed the Rock of his salvation." But it will never be so again. He who had come to conquer their foes, had come also to subdue their hearts. He "will pour upon them the Spirit of grace and of supplications, and they shall look upon Him whom they pierced;" for He will stand before them, as another Joseph in the midst of his astonished brethren, "and

F

they shall mourn for Him as one who mourneth for his only son, and shall be in bitterness for Him, as one that is in bitterness for his first-born." He will desire that they should mourn. Finally, indeed, He will dry their tears: but mourning is first to have its course. "In that day shall there be a great mourning in Jerusalem. The Land shall mourn, every family apart." It will still be true that the contrite in heart, and the poor in spirit, are alone blessed. And seeing that this rescued and forgiven people are destined for power, and are to be made "princes throughout all the earth," and that nations are to be regulated by them, it is meet that they who are to be entrusted with such power should themselves first be well broken in spirit; for what more terrible than power wielded by the proud, unbroken spirit of man. The hundred and thirty-first Psalm supplies the utterance of Israel then. "Lord, my heart is not haughty, nor mine eyes lofty: neither do I exercise myself in great matters, or in things too high for me. Surely I have behaved and quieted myself, as a child that is weaned of his mother: my soul is even as a weaned child. Let Israel hope in the Lord from henceforth and for ever."

Their day of conquest, therefore, ends in contrition and mourning. But grace will accomplish its work. Their subdued souls shall be brought into full acquaintance with the Fountain opened for sin and for uncleanness. "In that day there shall be a fountain opened to the house of David, and to the inhabitants of Jerusalem, for sin and for uncleanness." They will not only be sprinkled and forgiven through the blood of the Lamb, but they will understand also the reason and ground of their forgiveness, and like ourselves (who in this, forestall their blessings) will be able to say that their garments are washed and made white in the blood of the Lamb. Although regenerate, their flesh will yet remain, and in the flesh "good doth not dwell." They will need, therefore, from day to day, refreshment and consolation in recurrence to that blood once offered, which cleanseth from all sin. The Holy Spirit will abundantly be poured upon them, both that they might know the things that have been freely given to them of God, and also that they might be heralds of His salvation to the dark heathen world. They will be sent "to Tarshish, Pul, and Lud, that draw the bow, to Tubal and Javan, to the isles afar off which have not heard His fame, nor seen His glory, and they shall declare His glory among the Gentiles." (Isaiah lxvi. 19.)

Their own Land also will need cleansing, and it will gradually be cleansed. Many an unclean spirit will have been acting there; many a lying prophet will have prophesied; but then the hand of the Lord will be turned upon them in grace, to purge away the evil. And if any should venture yet to prophesy in the name of the Lord falsely, even his own parents who begat him, will be willing, rather than the Lord should be dishonoured, to resign their son to punishment, or even to death. "His father and his mother that begat him shall say unto him, Thou shalt not live; for thou speakest lies in the name of the Lord, and his father and his mother shall thrust him through when he prophesieth." The claims of nature and of natural love were once dominant in Israel. But they will cease to be so; then grace will have turned the course of the heart's deepest feelings, and they will refuse to flow, save only for God. And the prophets themselves, even they who have worn a rough garment to deceive, they who have purchased influence by self-imposed austerities—they too shall be reached, for grace is able to save, even to the uttermost. They shall be reached and changed, and the desire even of distinction shall cease to sway their spirits. The dignity of office will be surrendered by their repentant hearts. Each will say, " I am no prophet, but only a husbandman—one who was taught to keep cattle from my youth :" and he will willingly fall back into the sphere from which Satan, and not God, had raised him. He will bear also, and be willing to acknowledge the tokens of his former shame, and confess to the wounds which his deception had earned, from the hands even of his own friends. Here, indeed, is an instance of the subduing, sanctifying power of grace. It is hard to resign what we have prized, but harder still, after the resignation, meekly to bear the tokens of the shame. But there will be power of grace in Israel then; and it seems to be the object of this passage to record the completeness of its triumph over the hearts and ways of God's recovered people.

And if it be asked, how such grace could be shown to such a people—how those so distant in evil should suddenly receive such deliverance, and not deliverance merely, but strength; and how, after being so strengthened, they should be endowed with such riches of inward grace, the answer is this—that long before, there had been One of whom Jehovah had said, "Awake, O sword, against my shepherd, against the man that is my fellow, saith the Lord of hosts." That sword *had* awaked; the Shepherd *had been* smitten; and therefore this power and grace can be extended even towards that people who had trampled down the Shepherd, and rejoiced at beholding the scattering of the flock. It is on the atonement once made, that these marvellous actings of grace towards Israel are *grounded;* and hence this reference to the smiting of the Shepherd, after the detail of the forgiveness, restoration, and favour, that had been purchased by His sacrifice.

When the sword awoke against the Shepherd, the sheep were indeed scattered. All the disciples in terror forsook Him, and fled. Yet the Lord remembered them in grace, and turned His gentle hand upon the little ones. Here is our present blessing—the blessing of the feeblest who believes.

In the land of Israel, there is desolation, and a still severer doom stands yet pronounced against it. "It shall come to pass, that in all the Land, saith the Lord, two parts therein shall be cut off and die." But whilst the action of destructive judgment is thus towards Israel, there are some from among the Gentiles, and a few from among Israel, who believe on His name, and on them His hand is turned to guide, to strengthen, and to feed. And so it will be till the last of these sheep shall have been gathered in, and then He will remember Israel again, and though He will bring fires upon them, even fires that shall burn unto destruction against all but a remnant; yet that remnant, though it be but a third, shall be spared and blessed; "I will bring the third part through the fire, and will refine them as silver is refined, and will try them as gold is tried: they shall call upon My name, and I will hear them: I will say, It is My people; and they shall say, The Lord is my God."

CHAPTER V.

ON ZECHARIAH XIV.

I MAY now safely appeal to any who have seriously weighed the evidence of the preceding chapters, and ask them to say, whether certain great substantive facts touching the future, are not conclusively proved by them? It is proved that vast Gentile nations will again be gathered in siege against Jerusalem; that these hosts are in the land of Israel destroyed; that the Heads of Israel in Jerusalem are delivered and also converted; that consequently they must have returned to their land and city unconverted; and that they are delivered and converted by the personal manifestation of the Lord. These and other such events are ever recorded as the great characteristic features of that period which is termed in Scripture, "the end of this age."

There is only one of these facts that I can suppose as at all likely to be questioned, and that is the personal manifestation of the Lord. If the words, "they shall look upon Me whom they pierced," and again, "in that day the house of David shall be as God," are not considered conclusive, yet surely the fourteenth chapter, which we are now about to consider, must remove all ambiguity. "Then shall the Lord go forth, and fight against those nations, as when He fought in the day of battle, and His feet shall stand in that day on the Mount of Olives, which is before Jerusalem on the east; * * * the Lord my God shall come, and all the saints with Thee."

It is thus that one part of Scripture confirms and renders more definite the testimony of another. But the fourteenth of Zechariah is not merely confirmatory. It repeats, indeed, many of the statements that former chapters had made, but it enlarges also, and adds new features.

In the twelfth chapter our minds have been exclusively directed to the interference of the Lord *on behalf of* Jerusalem. In that chapter destruction is spoken of as being not against *her*, but against her enemies; and if it were not for the concluding words of the thirteenth, we might have almost supposed that no outpouring of judgment on Jerusalem, either immediately preceded or accompanied the day of her final visitation. But the fourteenth chapter supplies this deficiency. The first two verses lead us back to a period immediately preceding that with which the twelfth chapter opens, and speaks of a blow that had just been allowed to fall on Jerusalem, by means of these very nations, whose destruction the self-same chapter records. It is with this that the fourteenth chapter commences: "Behold a day cometh for Jehovah,* and thy spoil shall be divided in the midst of thee. For I will gather all nations against Jerusalem to battle; and the city shall be taken, and the houses rifled, and the women ravished; and half of the city shall go forth into captivity, and the residue of the people shall not be cut off from the city."

Such will be in part the instrumentality by which the Lord, even up to the very end, continues to punish that city: "You only have I known of all the families of the earth; therefore will I punish you for all your iniquities." (Amos iii. 2.) The Scriptures again and again speak of the wasting destruction that shall fall upon Israel until only a "remnant" of them shall be left. "Though Thy people Israel be as the sand of the sea, yet [only] a remnant of them shall return: the consumption decreed shall overflow with righteousness; for the Lord God of hosts shall make a consumption, even determined, in the midst of all the land." (Isaiah x. 22.) The assault and triumph of these Gentile nations will be one of the means by which this "decreed consumption" shall be wrought.

The nations will be allowed to complete their

* It is not the same expression as "the day of the Lord." When the latter expression is used, it always, I believe, denotes THE day in which Jehovah personally interferes, and does not include a prolonged period preceding.

capture, and even to carry half of the people into captivity. But after they have thus accomplished their purpose, and think themselves secure of the subjection of their foe, suddenly, for some untold reason (untold, at least, in this chapter), Jerusalem again excites their enmity, and again they gather together against her; but it is for the last time. The appointed hour for the interference of the Lord will have come, and He will accomplish that which He has promised in the twelfth chapter, and in the second verse of the chapter we are considering. "He will go forth and fight against those nations, as when He fought in the day of battle. And His feet shall stand in that day upon the Mount of Olives." From this verse onward the fourteenth chapter unites with the twelfth, and, like it, testifies to the pardon and deliverance of Jerusalem.

But Israel, as a whole, will be little prepared for this sudden visitation. They, as well as the nations, will be overtaken by it as by a snare. "Ye shall flee—yea, ye shall flee, like as ye fled from before the earthquake in the days of Uzziah, king of Judah." "Enemies," "adversaries," and "hypocrites" are spoken of as in the midst of Israel, even up to the hour of its final visitation. "The sinners in Zion are afraid; fearfulness hath surprised the hypocrites. Who among us shall dwell with the devouring fire? who among us shall dwell with everlasting burnings?" (Isa. xxxiii. 14.)

"Therefore, saith the Lord, the Lord of hosts, the mighty One of Israel, Ah, I will ease Me of Mine adversaries, and avenge Me of Mine enemies, and I will turn My hand upon thee, and purely purge away thy dross, and take away all thy tin; and I will restore thy judges as at the first, and thy counsellors as at the beginning; afterward thou shalt be called, The city of righteousness, the faithful city. Zion shall be redeemed with judgment, and her converts with righteousness, and the destruction of the transgressors and of the sinners shall be together, and they that forsake the Lord shall be consumed." (Isa. i. 24—28.)

The day of Jerusalem's deliverance is ever spoken of in the Scriptures as a day also of consuming judgment, so that only a remnant will be left; but that remnant, when delivered and strengthened, is delivered and strengthened *as a nation*. They are rescued, not as individuals merely, but nationally. "A nation shall be born in a day;" "A small one shall become a strong nation;" "*Israel* shall do valiantly"—are all texts referring to this period of their history, and speak of them as corporately and nationally delivered.

And as regards the means of their deliverance. It is said that "the Lord shall go forth and fight against those nations, as when He fought in the day of battle; and His feet shall stand in that day upon the Mount of Olives." Can any language be more simple or more definite than this? We believe (unless, indeed, we reject all the past testimonies of the Word of God) that there have been occasions of old when He has interfered in visible, almighty power on behalf of Israel, and delivered them. He divided the sea; He caused the walls of Jericho to fall down; He fought for them against the kings of Canaan; He descended on Sinai in their sight, when Sinai trembled and was shaken; and He has said, "yet *once more* I shake not the earth only, but also heaven." And if we believe that Jesus is Jehovah, and that His feet have already stood upon the Mount of Olives; that thence He departed from the earth when the angels said, "This same Jesus shall so come even as ye have seen Him go;"—why should we doubt that He will also stand there in glory? Why should we doubt what is so plainly written, that the Mount of Olives shall tremble and cleave, and bear witness thus to the presence of God? "Jehovah, my God, shall come, and all the saints with Thee." They will be with Him, because they will have met Him in the air, and will return, so as to surround Him and be with Him, when He is thus "revealed with His mighty angels, in flaming fire, taking vengeance."

It will be a day the most momentous of all days in the earth's history. It will not destroy the earth, for it is said immediately afterwards, that "the Lord shall be King over all the earth; in that day shall there be one Lord, and His name one;" neither will it destroy Jerusalem, for it is said, (ver. 10,) "Jerusalem shall be lifted up and inhabited in her place;" and again, "Jerusalem shall be safely inhabited." Many a heathen nation also will be spared, as we find from Isaiah.* But it will be a day the like to which has never been. "Alas! for that day is great, it is even the time of Jacob's trouble, but he shall be saved out of it." (Jer. xxx. 7.) It is emphatically called in the Old Testament, THE day of the Lord.

"It shall be one day known unto the Lord, not day nor night." It shall not be day, for all the natural sources of light shall be withdrawn: "the sun shall be darkened, in his going forth, and the moon shall not cause her light to shine." "The sun and the moon shall be darkened, and the stars shall withdraw their shining." (Joel ii.) On the earth, therefore, blackness of darkness will rest, such as was on the face of the waters before the Spirit of God moved thereon—before God said, "Let there be light." But in the midst of this abyss of darkness, there will suddenly be displayed the light of a glory too terrible for human eyes to behold. "He will come in His own glory, and in His Father's glory, and in the glory of the holy

---

* See Isaiah lxvi. 19. "I will send those that escape of them unto the nations, to Tarshish, Pul, and Lud, that draw the bow, to Tubal and Javan, to the isles afar off, that have not heard My fame, neither have seen My glory; and they shall declare My glory among the Gentiles."

angels." Saints, also, as well as angels, in all the brightness of their unearthly glory, will surround Him in myriads unnumbered. "A fire goeth before Him, and burneth up His enemies round about. His lightnings enlightened the world; the earth saw, and trembled." (Ps. xcvii. 3, 4.)

But it shall be only *one* day. The covenant with Noah forbids that this awful interruption of the course of nature should extend beyond *one* day: "While the earth lasteth, day and night shall not cease." Therefore night shall again resume its wonted course. The tempest of wrath shall have rolled over Israel and the earth, and will have wrought its work. It will have destroyed, not the earth, but it will have "destroyed them that destroy the earth." (See Rev. xi. 18.) The evening shall return with light, pure, serene, and blessed. The stars will again shine peacefully, and a delivered earth wait for the arising of a "morning without clouds."

We can, in part, conceive the feelings with which the spared remnant of Israel will behold the light of that evening—the evening which is to introduce the new order of God. They have been described in the twelfth chapter as subdued, contrite, and mourning. And no marvel: carried as they will have been, by a power that they knew not, through such a day of terror, strengthened for the Lord in it, and left at last in a scene of tranquil blessing, received from the hands of One whom they *had* despised, but to whom they have now learned to say, "My Lord, and my God;" it would be strange indeed if they should not, whilst numbering such mercies, be bowed in contrition of spirit. And when they shall at last be comforted, and the Spirit be poured out upon them from on high, when the knowledge of their own past history, and of the world's history, and of the Church's history, will all be opened to them in the light of God, then, like so many Pauls, monuments of sovereign grace, they shall go forth to the dark places of the earth, rich in experience and in the knowledge of God, and from them shall flow rivers of living water.

We read in many parts of the Scripture that the land of Israel will, in that day, teem with evidences of the miraculous power of God in dispensing blessings. On the sides of Zion, for example, the wolf and the lamb, the leopard and the kid, shall be seen together, and a little child shall lead them. Nothing shall hurt or destroy throughout God's holy mountain. These will be sights that no one will deny to be in themselves blessed. But they are symbols, also — living symbols, speaking of higher blessings: for they indicate the peace and harmony and love that shall pervade all hearts and all peoples whom the power of Zion shall effectually reach. And if God has appointed that the spiritual influence of which I have spoken above should go forth from His forgiven and privileged nation in Jerusalem, we might expect to find some outward symbol of this, its relation. And, accordingly, a symbol is given in the perennial flow of those streams which, going forth from the sanctuary in Jerusalem, shall heal waters which, like the Dead Sea, have been accursed, and spread life and refreshment in the midst of desolation. "It shall come to pass in that day that living waters shall go forth from Jerusalem, half of them toward the former sea and half of them toward the hinder sea, in summer and winter shall it be." Ezekiel, in vision, saw them issue forth as "a river that he could not pass over, for the waters were risen, waters to swim in, a river that could not be passed over. And he said unto me, Son of Man, hast thou seen this? Then he brought me, and caused me to return to the brink of the river. Now, when I had returned, behold, at the bank of the river were very many trees on the one side and on the other. Then said he unto me, These waters issue out toward the east country, and go down into the desert, and go into the sea; which being brought forth into the sea, the waters shall be healed. And it shall come to pass, that every thing that liveth, which moveth, whithersoever the rivers shall come, shall live:" etc. (Ezekiel xlvii. 6—9.)

No one, I suppose, who reads Ezekiel, especially his reference to En-gedi and En-eglaim, will doubt the literality of the fact. Yet we should not, on that account, forget the more blessed spiritual relation of Zion and Jerusalem which this fact symbolises—when "out of Zion shall go forth the law, and the word of the Lord from Jerusalem."

National supremacy, *i.e.*, supremacy in the control and regulation of nations, as well as supremacy in religious truth, will attach to Jerusalem then. Diffusion of truth, such truth as saves and enlightens individual souls, will, doubtless, be the most important function of that favoured city. It will be diffused partly by the service of its people, severally; partly by its own corporate testimony. Individually, the saints in Israel will not be deprived of that which is the joy and comfort of the saints now, even to preach and to teach the Lord Jesus Christ, and to cherish others with that same kind of care with which the Apostle once sought to nurture the Churches—"gentle among them, even as a nurse cherisheth her children;" for though Satan will be bound, yet the flesh in which no good thing dwelleth, will still remain, so that gracious, kindly, shepherd care will continue to be needed. Jerusalem will doubtless esteem it her highest calling to be the pillar and ground of the truth; the golden candlestick fed with golden oil. (See Zech. iv.) But as the chief of nations, also, it will exercise authoritative control over the earth—a control most blessed in its exercise, because exerted for, and in subordination to, TRUTH. Then at last there will be no contrariety, no collision between the ministration of God's truth and the sovereignty of the Throne; for He who sits upon the Throne as the King of kings will also be the Priest of the Most High God. "He will sit as a Priest upon His Throne," He will be the true Melchizedek, King of Righteousness and King of Peace; but Priest also of the Most High God, Possessor of heaven and earth. The supremacy, therefore, of Jerusalem and its King, the nations will be required to recognise. One prescribed test of their obedience will be their coming up to Jerusalem year by year to keep the Feast of Tabernacles.

The Feast of Tabernacles was the most joyful of all the feasts of Israel. Throughout it they are commanded to dwell in booths, in remembrance of the time when they knew the sorrows of journeying through a waste and howling wilderness. But the wilderness will no longer be around them. They will be in the land of their long-promised blessing: the joy of all lands. "Thou shalt no more be termed Forsaken—neither shall thy land any more be termed Desolate; but thou shalt be called Hephzibah, and thy land Beulah, for the Lord delighteth in thee, and thy land shall be married." (Isa. lxii. 4.) And seeing that nothing so much as contrast gives liveliness to our apprehensions, and that nothing so much heightens joy as the remembrance of sorrow *past*, there will be brought before the recollections of Israel, by the booths in which they sojourn, the features of a former scene, that will stand in strange, but blessed, contrast with their then present and abiding joy. Around them will be the ingathered riches of their Land; pledges and tokens of their being recipients not only of spiritual, but also of all natural, blessings from the hand of the Lord their God. "Thou shalt observe the Feast of Tabernacles seven days, after that thou hast gathered in thy corn and thy wine, and thou shalt rejoice in this feast * * * because the Lord thy God shall bless thee in all thine increase, and in all the works of thine hands, therefore thou shalt surely rejoice." (Deut. xvi.) And in that day they *will* rejoice; "When the Lord turned again the captivity of Zion, we were like them that dream. Then was our mouth filled with laughter and our tongue with singing; then said they among the Gentiles, The Lord hath done great things for them. The Lord *hath* done great things for us, whereof we are glad." (Psalm cxxvi. 1—4.)

There seems a peculiar suitability in the Gentiles being called to witness this great feast of the joy of Israel. They had been the witnesses of its desolation, and many of them had been instrumental in causing that desolation. As fierce and cruel monsters ("beasts," θηρια), they had devoured God's heritage. They had trodden it down, and exulted in its degradation. But now they will be gathered to behold its glory and its joy. "The Gentiles shall come to thy light, and kings to the brightness of thy rising." (Isa. lx.) But they shall not merely behold; they shall also share the blessings of Israel. Grace will lead even to this; for grace loves to widen the streams of its goodness. And, therefore, when we read of Israel's blessings, we read also of the Gentiles being called on to give thanks. "Rejoice, ye Gentiles, with His people." "Sing unto the Lord a new song, and His praise from the end of the earth, ye that go down to the sea, and all that is therein; the isles, and the inhabitants thereof. Let the wilderness and the cities thereof lift up their voice, the villages that Kedar doth inhabit: let the inhabitants of the rock sing: let them shout from the top of the mountains." (Isa. xlii. 10, 11.) These and such like passages plainly show that the inhabitants of the world at large are called to share in the joy of Israel in that day.*

* The Feast of Tabernacles was the only one of the feasts of Israel that had an eighth day. (Lev. xxiii. 39.) The eighth, being the day after the Sabbath—the day on which the Lord rose from the dead, is always typical of resurrection-life in the new creation. Israel, in the Millennium, will be reminded by it, that their final rest is not in the fair scene of prosperity which is then spread around them. They will look beyond, even to the "new heavens and new earth," wherein dwelleth righteousness. In the Millennium, Israel will be yet in the flesh, and "in the flesh good doth not dwell." "The flesh lusteth against the Spirit." It baits and

Yet, in the midst of all this prosperity, and glory, and joy, Jerusalem will remain holy. Their "heart shall not become uplifted, nor their tongue haughty." Other nations, even while the hand of

allures away from God and His ways, even where Satan is not, to stimulate its enmity. Their bodies also will be unredeemed, and the earth, though renovated, will not be *new*. The Lord will not yet have said, "Behold, I make all things new." Restraint of moral evil, or repression of natural decay, through the power of God in redemption, is blessed: yet it is very different from a condition in which there is nothing to be repressed, because all is good. This last condition will not pertain to Israel and the nations of the Millennium, until the Millennial earth and heavens have passed away, and no place been found for them.

In the Millennium certain distinctions connected with the flesh, and which have no place in resurrection, will still continue, such, for example, as those between male and female, Jew and Gentile. In the Millennium, the Gentiles, though entirely one with the Jews in all *spiritual* privileges, (for in spiritual blessings there is no difference between Jew or Greek, male or female,) yet, as to national position, will stand in a secondary place. "The first dominion will come to the daughter of Jerusalem." (Micah iv. 8.) This seems, therefore, to be another reason why the Gentiles should come up to celebrate that feast to which the eighth day was added; for it spoke both to Israel and to them, of a future and more blessed hour, when there should be perfect co-equality in heavenly blessing in the new creation.

That typical ordinances, such as the celebration of the Feast of Tabernacles, are not inconsistent with true spiritual religion, is proved by the appointment of two symbolic rites now—Baptism and the Lord's Supper. If such symbolic

the Lord is pouring blessings upon them, will be discontented and disobey. Some will refuse to go up to keep the Feast of Tabernacles, and will receive, therefore, the due chastisement for their rebellion. But it shall not be so with Israel. The same grace that has given them their unequalled blessings, will have also subdued their hearts to know and to fear the Lord their God. The prayer which they had offered will have been heard. They had said: "Let Thy hand be upon the man of Thy right hand, upon the Son of Man, whom Thou madest strong for Thyself. *So will we not go back from Thee.*" (Ps. lxxx. 17, 18.) And they will not go back from Him: "In that day shall there be upon the bells of the horses holiness unto the Lord;" that is, the whole external character of life (for that it is which is exhibited in the streets of a city) shall bear in all its parts, throughout all its detail, the impress of holiness unto the Lord. Religious life and fellowship shall be holy also; for the pots in the Lord's house, vessels which of old the priests had so often defiled, shall be like the bowls before the altar—holy. Private and domestic life shall be hallowed too; for "every pot in Jerusalem and in Judah shall be holiness unto the Lord of hosts, and all they that sacrifice shall

---

rites do not militate against spiritual worship now, much less need the Feast of Tabernacles be objected to then, seeing that it is so peculiarly connected with the *national* order of God's government.

come and take of them, and seethe therein; and in that day there shall be no more the Trafficker in the house of the Lord of hosts."*

* The word which I have translated "Trafficker," is the same as Canaanite. It seems to be an allusion to that which had given its characteristic feature to the closing period of the Times of the Gentiles, viz., mercantile power. See Zechariah v.—vision of the Ephah, and the whole of the eighteenth of Revelation, as considered in "Babylon: its Revival and Final Desolation," as advertised at end of this volume. See also "Prospects of the Ten Kingdoms," page 88.

CHAPTER VI.

FUTURITY OF THE MANIFESTATION OF ANTI-
CHRIST—HIS CONNEXION WITH JERUSALEM.

THERE are few things more necessary in prophetic enquiry, than to mark the place which JERUSALEM occupies, in the dispensational arrangements of God. The chapters which we have just been considering in Zechariah, sufficiently manifest the wonderful character of the events yet to occur in that city. A city on behalf of which the Lord will visibly interfere, and which He will so marvellously strengthen—a city that is to be so marked by holiness, and destined to be the centre whence light and truth are to be diffused over all nations, must be important in the sight of God, in proportion as He values that truth, and the spread of the knowledge of His own holy name. Ethiopia shall not stretch out her hands unto God, until Israel shall first have become white as snow in Salmon. (See Ps. lxviii.) The ends of the earth will not fear Him, until after He has lifted up the light of His countenance on Israel. "Jehovah" (I quote the words of one of Israel's Psalms, lxvii.) "Jehovah shall bless US," and afterward "all the ends of the earth shall fear Him." Well, therefore, may it be said, "Pray for the peace of Jerusalem: they shall prosper that love thee." (Ps. cxxii. 6.) "Ye that make mention of the Lord, keep not silence, and give Him no rest, till He establish, and till He make Jerusalem a praise in the earth." (Is. lxii. 6.)

After the Flood there were great developments of energy in many Gentile nations. Assyria under Nimrod, and Egypt under the Pharaohs, early rose into eminence. Both Egypt and Assyria were restrained, until it had been fully proved whether or not Jerusalem was worthy of supremacy in the earth. No Gentile empire was allowed to rise into universal sovereignty—all were held, as it were, in check, until Israel and Jerusalem had been sufficiently tried, and had fully manifested their unworthiness. This is one instance in which the dealings of God with the nations were made dependent on the condition of Jerusalem.

And when Israel and its kings had by transgression lost their blessings, and the glory of the reign of Solomon had faded away, the supremacy, which was taken from them, was given to certain Gentile nations, who were successively to arise and bear rule in the earth, during the whole period of Israel's rejection. The first of these was the Chaldean empire under Nebuchadnezzar. The period termed by our Lord the "Times of the Gentiles," commences with the capture of Jerusalem by Nebuchadnezzar. It is a period, coincident from its beginning to its close, with the treading down of Jerusalem. "Jerusalem shall be trodden down of the Gentiles till the Times of the Gentiles be fulfilled." Nebuchadnezzar therefore, and the Gentile empires that have succeeded him, have only received their pre-eminence in consequence of Jerusalem's sin; and the reason why they were endowed with that pre-eminence was, first, that they might inflict the appointed chastisements on Jerusalem; and secondly, that they might be made a governmental centre to the earth during the time that Jerusalem is being chastened. When they shall have fulfilled these purposes, they shall themselves be smitten, because of their own evil, and be made "like the chaff of the summer threshing-floors." In this we have another evidence that the earthly dispensations of God revolve around Jerusalem as their centre.

The method which it has pleased God to adopt in giving the prophetic history of these nations, is in strict accordance with this principle. As soon as they arose into supremacy and supplanted Jerusalem, Prophets were commissioned, especially Daniel, to delineate their course. We might perhaps have expected that their history would have been given minutely and consecutively from its beginning to its close. But instead of this, it is only given in its connexion with Jerusalem: and as soon as Jerusalem was finally crushed by the Romans, and ceased to retain a national position, all detailed history of the Gentile empires is suspended. Many a personage, great and glorious in the world's history, has since appeared. Mahomet and Saladin, Charlemagne and Napoleon have arisen; mighty battles have been fought; kingdoms raised; and kingdoms subverted; yet Scripture passes silently over all such events (however important in the annals of the Gentiles) because Jerusalem (having *nationally* ceased to be) is *nationally* unaffected. Eighteen hundred years ago, when Jerusalem was laid desolate, all detailed history respecting the Gentile nations was suspended. It is suspended still; nor will it be resumed until Jerusalem re-assumes a national position. Then the history of the Gentiles in its connexion with Jerusalem is again minutely given; and the glory and dominion of their last great king described. His history will be especially connected *with Jerusalem and the Land*. "He is to glorify himself *on Zion* (Isaiah xiv. 13; Daniel xi. 45), and to be broken and trodden under foot *in the Land, and on the mountains of Israel*." (Isaiah xiv. 24—27.)

That the Jews whilst yet unconverted will go back to their Land, and there re-assume a national standing, has been already proved from the concluding chapters of Zechariah; for if, as those chapters teach, they are nationally converted *when in their city*, they must of course have returned to it when unconverted. Nor would any one who contemplates their present circumstances in the light of mere human probability, esteem that return to be improbable. Their land is, as it were, waiting for them—unoccupied. It is still a goodly land, and as forming the great link of connection between Western Europe and the Euphratean countries, it is peculiarly adapted for an intellectual, enterprising and wealthy people, whose idol is commercial gain. The Gentile nations also, who used to persecute have begun to favour them. "Many," says the prophet Daniel, when speaking of this period, "shall cleave to them with flatteries." (Chap. xi. 34.) Men are beginning to speak of them as an ancient people, a people wise in their generation and energetic, useful therefore for purposes of national advancement and prosperity, and necessary for the pacification of the East. What wonder then, that the combined influences arising from these, and other like things, should lead them back to their city, and that the blindness of the present hour should be ready to mistake that return, for their restoration under the hand of the Lord in blessing?

The condition of their Land when they return will be just such as might be expected from their character. "Replenished from the east * * * * their land will be full of silver and gold, neither is there any end of their treasures; their land is also full of horses, neither is there any end of their chariots; their land also is full of idols." (Isa. ii.) Such will be the aspect of their land; and as to themselves, pride and haughtiness, the loftiness of the cedar of Lebanon, and the stiff sturdiness of the oak of Bashan will be their characteristics, even till the day of the Lord shall come. There is an awful passage in Ezekiel that speaks of this their return. (Chap. xxii. 19—22.) "Therefore thus saith the Lord God, Because ye are all become dross, behold, therefore I will gather you into the midst of Jerusalem. As they gather silver, and brass, and iron, and lead, and tin, into the midst of the furnace, to blow the fire upon it, to melt it; so will I gather you in my anger and in my fury, and I will leave you there, and melt you. Yea, I will gather you, and blow upon you in the fire of My wrath, and ye shall be melted in the midst thereof; as silver is melted in the midst of the furnace, so shall ye be melted in the midst thereof; and ye shall know that I the Lord have poured out My fury upon you."

The last great king that is to arise among the Gentiles—one who is called in Daniel emphatically, "THE King" (xi. 36) and "THE Prince that shall come," is one of the chief instruments by which the judgments of the Divine hand will reach that evil people. As the history of this person is so important, and leads us to the very climax of Jewish and Gentile iniquity, it is needful that the statements of Scripture respecting him, should, in outline at least, be familiar to our minds. If his rise be future, it behoves us to *know* that it is future. My present object therefore will be to quote such passages as prove his futurity, and describe his actings in relation to Jerusalem. The first chapters to which I will refer are the eighth and the concluding part of the eleventh of Daniel.

The subject of the Book of Daniel as a whole, is the indignation of God directed through the instrumentality of the Gentile empires upon Jerusalem: but the vision of the eighth chapter is expressly said to relate *to the end of that indignation*. This is twice stated in this chapter; first, in the seventeenth verse:

"Understand, O son of man, for AT THE TIME OF THE END shall be the vision."

And again in the nineteenth verse:

"Behold I will make thee know what shall be in the LAST END OF THE INDIGNATION; for at the time appointed the END shall be."

Unless therefore we are prepared to say that the indignation against Jerusalem has terminated, and that the time of blessing, both on Israel and on the nations, has come; we must admit that the king described in this chapter, has not yet arisen. Facts prove that no such person is at present acting in the earth; and he cannot have finished his course, and be no more, for in that case Jerusalem would have been forgiven, and the indignation against her have ceased.

Whenever this last great King of the Gentiles does arise, he will be known by answering to the criteria supplied by this chapter, which are as follow:

I. He must arise, not from the West, but from part of the EASTERN division of the Roman Empire, viz., from one of the four parts into which the empire of Alexander the Great was divided. This we are taught in the eighth verse. "Therefore the he-goat waxed very great; and when he was strong, the great horn [which is said in verse 21 to be the first king of Greece, Alexander] was broken; and for it came up four notable ones toward the four winds of heaven. And OUT OF ONE OF THEM came forth a little horn which waxed exceeding great towards the South, and toward the East, and toward the pleasant land." See also the 21st and following verses.

II. He will find the daily sacrifice offered at Jerusalem, and will take it away. This is taught in the eleventh verse. "Yea, he magnified himself even to the prince of the host, and by him the daily sacrifice was taken away, and the place of his sanctuary was cast down."

III. He is to arise "when the transgressors are come to the full." (Verse 23.)

IV. He is to act during the last end of the indignation against Jerusalem. (See verses 17 and 19, already quoted.)

V. He exalts himself not only against Jehovah as Jehovah, but also against the Prince of princes, *i.e.* against the Lord Jesus, the Messiah of Israel.

VI. He is broken without hand, that is, by no human instrumentality, but immediately by the Lord. This is taught us in the 25th verse. "He shall also stand up against the Prince of princes; but he shall be broken without hand."

It is scarcely needful to say, that no one, in whom these characteristics are united, has yet arisen. Neither the Pope, nor Mahomet, nor Antiochus Epiphanes (all of whom have been supposed to fulfil this chapter) answer to these tests. Mahomet and the Pope answer to none. Mahomet arose from Arabia, which according to God's covenant with Ishmael, has always remained independent of the four great Empires that have successively arisen, and has never been incorporated with either. Antiochus, however, went far towards fulfilling the predictions of this chapter, and was doubtless intended in an especial manner to prefigure Antichrist. Antiochus visited Jerusalem; took away the daily sacrifice and profaned the Temple; (see first chapter of first book of Maccabees) but he did not live in "*the last end of the indignation;*" nor "*when the transgressors are come to the full;*" nor "*in the latter time of those kingdoms*" (for Greece and Egypt are still ruling); nor did "*he stand up against the Prince of princes*" (for Christ had not then been manifested); nor was he "*broken without hand*"; nor did Antiochus Epiphanes spring "*out of*" one of the four; for he was *one of the four;* nor was he "a little horn," that is, a little Potentate; he was a horn emphatically "great." The fulfilment therefore of this chapter must be future.*

Let us now turn to the eleventh chapter. That chapter after mentioning, in the 33rd verse, the captivity of the Jews under the Romans, passes rapidly and without detail (in accordance with the principle to which I have above referred) over the present lengthened period of Jerusalem's desolation, and hastening on to the "time of the end," renews its *detailed* history as soon as he, who is emphatically called "THE KING," appears upon the scene. "THE KING" shall do according to his will; and he shall exalt himself, and magnify himself above every god, and shall speak marvellous things against the God of gods, and shall

---

* Antiochus Epiphanes was doubtless a remarkable type of Antichrist, and was generally so regarded by the early Christian writers. Sulpitius Severus, born A.D. 366, says, that his desire to establish uniformity led him not to persecute the Jews only, but the Gentiles too. " He (Antiochus) did not even spare his Gentile subjects. He tried to make them relinquish their inveterate superstitions, and to make them adopt one form of worship (*unum ritum*)." Lib. I., cap. xxi.

For further remarks on this subject see "Prospects of the Ten Kingdoms."

prosper till the indignation be accomplished: for that that is determined shall be done. * * * * And he shall plant the tabernacles of his palace between the seas in the glorious holy mountain; yet he shall come to his end, and none shall help him. * * * * And AT THAT TIME thy people (*i.e.*, Daniel's people—Israel) shall be delivered, every one that shall be found written in the book. And many of them that sleep in the dust of the earth shall awake, some to everlasting life, and some to shame and everlasting contempt. And they that be wise shall shine as the brightness of the firmament; and they that turn many to righteousness, as the stars for ever and ever."

Are these things fulfilled? Has there been one who has prospered till *the indignation against Jerusalem has been accomplished?* Has all "that determined" against Jerusalem been done? Has any one thus glorified himself on the glorious holy mountain? Has such an one been destroyed after having so glorified himself, and that, at the time of Israel's deliverance, and the resurrection of the saints? Is Israel delivered? Are the saints raised and shining as the brightness of the firmament? If, then, none of these things have yet come to pass, the prophecy respecting this Wicked King must still remain to be accomplished.

The seventh chapter of Daniel describes this same last monarch of the Gentiles under the symbol of "a little horn," the agent of the power of the Beast from which it springs—that Beast representing *all* the kingdoms which were of old included in the Roman Empire. This chapter is precise in fixing the period of this monarch's evil reign, as immediately antecedent to the Advent of the Lord in glory; and thereby identifies itself with the chapters we have already been considering. Thus in the eighth of Daniel, he is said to flourish "*in the last end of the indignation*"—in the ninth, "*until the consummation*"—in the eleventh, "*until the indignation be accomplished*"—in the seventh, "*until the Ancient of days came, and judgment was given to the saints of the high places; and the time came that the saints possessed the kingdom.*" "End of the indignation," "consummation," and the like, are all equivalent expressions, referring to the same period. If therefore the passages that have been quoted from the eighth, ninth, and eleventh chapters, are future, the prophecy respecting the Little Horn in the seventh chapter, must be future also. The Little Horn could not reign after the indignation against Israel had been accomplished.

Not that it is necessary to go from the seventh chapter itself in order to prove the futurity of its prediction respecting the Little Horn. That chapter teaches us that the whole of what was once the Roman Empire (that is, from Assyria to Britain) is to be divided into Ten kingdoms, and that these

Ten kingdoms, when once developed, are to continue until the close of the dispensation. Are there any such ten kingdoms dividing the Roman World at present? If not, there are no ten kings amongst or behind whom Antichrist *could* arise; or (to use the symbols of another chapter) no ten toes of the Image on which the Stone *could* fall and "grind them to powder." The very fact that there is, at present, no one who concentrates in his own person the power of the *whole* Roman World is of itself a sufficient evidence that the prediction is future. When once he has arisen, he will not cease to reign, and to persecute the saints throughout all the kingdoms over which he rules, until the Millennial kingdom comes. This is expressly taught in the 21st and 22nd verses of this chapter. "I beheld, and the same horn made war with the saints, and prevailed against them, UNTIL the Ancient of days came, and judgment was given to the saints of the high places (see Chaldee); and the time came that the saints possessed the kingdom." The present is a time of latitudinarian liberality, not of persecution. Even if the Western Branch of the Roman Empire were alone regarded, there is at present no Sovereign Individual, or System, that rules over it *all*—neither are there Ten kingdoms to be found within its scope.

Nor are the statements of the New Testament discordant with the Old. It would be impossible, if Daniel and the Revelation both describe (as they confessedly do) the period which immediately precedes the appearing of the Lord, that they should speak of the same Ten kingdoms, and of the same city—Jerusalem—without mentioning the great characteristic fact of that period, viz., the manifestation and reign of this great instrument of Satan.

Accordingly, we find the Revelation describing him in close accordance with Daniel, and using in some instances almost the same expressions:

- Daniel. "He shall speak great words against the Most High." (vii. 25.)
- Revelation. "Speaking great and blasphemous things." (xiii. 5.)

- Daniel. "He made war with the saints and prevailed." (vii. 21.)
- Revelation. "He made war with the saints and overcame them." (xiii. 7.)

- Daniel. "They shall be given into his hand until a time and times and the dividing of time," *i.e.*, 1260 days. (vii. 25.)
- Revelation. "Authority was given unto him to act forty and two months," *i.e.*, 1260 days. (xiii. 5.)

In the thirteenth chapter of the Revelation, the distinctive characteristics of Antichrist are given even more fully than in the chapters we have been considering. Each fresh characteristic as it is mentioned adds new proof that the person so characterised has not yet appeared. Who is there for example that is now "making war with the saints and overcoming them" throughout the WHOLE Roman World?

Who is there that controls the Ten diadems of the Roman World?

Where are there any such Ten diadems?

Who is there worshipped by every one whose name is not written in the Lamb's book of Life?

Who is there, who, himself distinctively a king, has standing in his presence, one who acts as an ecclesiastical minister on his behalf, and gathers around the King, or rather, around his Image, the worship of all over whom he reigns?

Where is there any such Image, or any such worship now?

Where is fire being made to come down from Heaven in the sight of men, or where is there an Image miraculously made to speak?

None of these things are seen as yet. THE "strong delusion" and THE "deceiving miracles," of which the Apostle speaks are not yet sent. But must not *a* strong delusion have fallen on *us*, if we believe that these things have been accomplished in a past 1260 years, or in any other past period? Do we not, if we assent that these things have occurred, or that they are occurring now, constitute ourselves "false witnesses against God," cata Θεου. Ignorance is no excuse; for ignorance, in the presence of such light as God has been pleased to vouchsafe on these subjects, must be self-caused, voluntary, ignorance. Could the Scriptures be "*searched*" and such ignorance continue?

The countries once included within the Roman Empire, are those which are to be placed under the headship of this great instrument of Satan. We cannot wonder that these countries should be judicially visited, and that "strong delusion should be sent on them to believe a lie," when we remember how great their responsibilities have been. In these countries the light of revelation, first through Israel, and afterwards through the Christian Church, has most peculiarly shone. In them, Christ and His Apostles personally ministered. Supreme power, also, in the earth, and consequent responsibility as to the use of such power, has from the time of Nebuchadnezzar to the present hour, rested with these nations. Here, arts and civilization have chiefly flourished; and the influence thus possessed has stamped, religiously and secularly, a character on the whole world. And what has that character been? It would be vain to argue, if conscience does not rightly answer that question.

The Eastern part of the Roman Empire was the birthplace of Christianity. Syria, Asia Minor, and the European parts of Turkey, abounded with Churches which the Apostles themselves watched over, and which for a time walked worthy of Him who had called them into His kingdom and glory. Antioch, Ephesus, Thessalonica, Corinth, as known in Scripture history, recall to our minds the freshness and power of Apostolic Christianity. In those, and in other cities, Churches were gathered worthy of being represented before God by "Candlesticks of gold"—lamps of the Sanctuary. But what are those regions now? Has the light by which they were thus visited been retained? No. Christianity itself became in those very cities corrupted; and now the hardened Apostasy of Mahomedanism, and the abominations of the Eastern Churches, as evil or more evil than those of Popery itself, have settled in upon the very countries which were once the home of the early and pure testimony of Christian Truth.

And if we turn to the Western division of the Roman Empire—speaking generally, it is divided between the superstitions and idolatries of Popery, and the rising power of democratic infidelity. Protestantism as exhibited in Switzerland, and many of the German provinces, has become the parent of heartless infidelity. In the midst of all these corruptions, stand, proudly isolated, the Jewish people—civilised, skilful, and energetic, but not less opposed to God and to His ways, than when they said of the Lord of glory, "Not this man, but Barabbas"—ignorant of their own present condition, and of their future prospects, yet scarcely more ignorant than the professing Christianity that surrounds them.

But whilst the energies of Truth have thus decayed, the vigour of the world under Satan, aided by debased and prostituted Christianity, is being developed to an extent never before equalled. Not only is the whole earth traversed, and its resources laid open to human enterprise; but the inventions of art, and the discoveries of science, have furnished men with so many new and mighty instruments of action, that a fresh era has confessedly been commenced in the history of mankind. The power thus acquired has already been directed in part, and will be directed yet more decidedly, towards the renovation of those countries, which were the birth-place and home of early civilisation; but which, because of their iniquity, have been smitten and brought low under the judgments of God. The condition of Nineveh, Babylon, Egypt, and Jerusalem, has borne witness to the severity of the blow. But when renovated by the influences of Western civilisation, will those countries be at all nearer to God and to His truth, than they were in times of old? If Egypt, and the cities of Asia Minor and Syria, are to be inoculated with the principles that prevail in France or Italy, and (I may now add) England, will their moral condition be better than in the days when Pharaoh reigned in Egypt, or Nebuchadnezzar in Babylon? And when we remember that the regeneration (as men will term it) of the East, is mainly to be effected by the return of the Jews hardened in guilt, to their own Land and City, we can easily understand how great must be the worldliness, and how flagrant the evil, of countries that are moulded under their hand. That also will be the hour when the Ephah in which "Wickedness" is hidden is to be transplanted to the Land of Shinar, and "set there upon her own base." (See Zechariah v.) *

If at such a moment the Lord Jesus were again presented to the world in the meek and lowly ministration of grace, the result would be the same as when He visited the earth before—He would again be despised, rejected and abhorred. The only difference would be, that the scorn would be greater and the hatred more intense, even as He said, "If they do these things in a green tree, what shall be done in the dry?" that is—if in the youth time of their evil, they reject and crucify Me, what will be done when the old age of matured iniquity shall have come?

The Holy One of God will indeed never again be subjected to the contumely and reproach of a rebellious world, except indeed as He still suffers in His people. The world has been tested by the mission of Him who came in His Father's name, and it has rejected Him. It is soon to be tested by another who shall come in his own name, glorifying himself; and him it will receive. He will be one in whom Satan will peculiarly dwell—"his

* See this subject considered in "Prospects of the Ten Kingdoms of the Roman World," as advertised at end.

coming," says the Apostle, "shall be after the inworking (ενεργειαν) of Satan."

Satan indeed will not, for he cannot, become man. It belongs to Omnipotence alone, to unite two distinct natures in one undivided Person. Jesus was not indwelt in by God merely, He *was* God manifest in the flesh—God and man in one Person. This coming instrument of Satan, will not be Satan manifest in the flesh. He will be simply a man; but a man in whom Satan will live and act, in a manner in which he has never lived and acted in man before. God will permit it, for the coming of Antichrist is a judicial infliction of His hand.

Satan has long studied and practised on the human heart. He knows how to dazzle men by the attractiveness of power, and how to cause them to quail before its terror. He can gratify the taste by refinement, and the understanding by knowledge; he can delight the ear with melody, and the eye by beauty; he can stimulate energy, direct enquiry, and employ his own knowledge of the secrets of creation in assisting men to subdue the powers of nature to their will—delighted to make men great if only he can control that greatness when attained, and use it against God.

Hitherto, indeed, his power has been restrained and often counteracted by the truth and testimonies of God. The brightest fires of Satan's kindling cannot but burn dimly whenever God sheds around them the power of heavenly light.

They require the full darkness of night in order to shine in the strength of their deceiving brightness. Hitherto light has been continually given from above, but a time is coming when it will be peculiarly withdrawn; and God will Himself send on men "strong delusion, that they should believe a lie."

If we examine the present condition of the nations that fall within the scope of the ancient Roman Empire, we find that all are more or less in a state of transition. Not only are the governments being remodelled, but, besides this, those forms of religion which have been hitherto dominant in particular countries are being so counterbalanced by the introduction of other influences as to be deprived of that exclusive pre-eminence which they have, in past ages, possessed. In Turkey, for example, and in Egypt many principles are being gradually introduced which will take from Mahomedanism exclusive dominance there. The reign of Popery in countries where it was once supreme is being checked by other influences. In our own country the increase of Popery, as well as the spread of latitudinarian infidelity, is taking from Protestantism the exclusive position it once occupied here. *Neutralisation* of religious influences anciently dominant, not their *destruction*, is the characteristic of the present hour. Thus it will become more easy for the secular power to assume the place of authoritative control and to mould the religious systems into subservient subjection.

All this might be well, if the secular power were indeed the servant of God's truth and appointed by Him, for the work of final rectification. But secular government among the Gentiles, never has been the servant of the Truth. As regards its present condition, the constitutional principle, or (as it has been termed) the principle of self-government, must necessarily make governments subservient to the will of the people. In former times, even after they began to acknowledge the people as the *source* of power, Governments professed to have a judgment of their own, and to rule according to that judgment. Now, they seek to embody the will of society—in other words, to embody and carry out the will of those, whom, professedly they govern, but whom in reality they serve. If Judaism be found among the people, Judaism must be honoured. However abominable the idolatries of India and the East, they must be upheld. A Temple must be maintained for Juggernaut; and the mysteries of Buddha reverenced. This has been habitually done by England in India, and in Ceylon. The Priests of Rome must be educated and sustained; Socinianism, if necessary, patronised or endowed; for society has a right to think for itself, and the mind of society is that which government is intended to express. Many, too, are beginning to speak as if God had not revealed His will; or revealed it so obscurely, that it cannot be understood. "Truth and error, (says a living* and influential statesman,) what words, what mockeries are these in the lips of such as we, and of all like us! Truth and error!—which perhaps may escape the accurate discernment even of angelic natures. I doubt even if we could summon before us some bright inhabitant of the upper sphere, whether he might not be the foremost to tell us, that the Almighty has made all His creatures to love Him, and none to comprehend Him." The inference from this (which is simply disguised Deism) is, that, if truth cannot be distinguished from error, Governments need not, indeed cannot, attempt to discriminate between them. It is assumed that the Scriptures are so obscure, that they cannot be appealed to as a guide. They will not help us to determine what Idolatry is, or whether it be, or be not, a sin. It is possible that the doctrines of Mahomet, and of Socinus, and of "the Jewish Teacher," may be only various aspects of truth, capable of being reconciled, if we had only sufficient penetration to discern the hidden links of unity. At all events, whatever be right, or whatever be wrong, the mind of society must be expressed; and the government, seeing that it owes to society its creation and its authority, is the fit organ for its expression. Such are the

* He has died since the second edition of this work was published.

growing principles of the present hour. Shall not all who fear God strive against them, and resist them, even, if needful, unto death? Shall they not cleave more closely than ever to the plain, clear, simple, instructions of God's Word, and show in their testimonies, and in their lives, that those instructions are plain and simple, and that they *can be* understood, explained, and acted on. God has not forgotten the preciousness of His truth, nor forsaken the plans of His own eternal wisdom. He will in due time visit these abominations; although for the present, He may allow them to progress, until they are in their maturity, and meet those vials of wrath of which the Book of Revelation warns.

At the present moment, men are not prepared for the blasphemies of *open* infidelity.* In some instances, as has been lately seen in France, such blasphemies are also connected with principles which shake the arrangements of society to their centre. But even where this is not the case, men are not yet prepared for an open rejection of the name and authority of God; and are therefore not ripe for the full development of Antichristianism. An intermediate link is necessary.

Accordingly, the system which is now beginning to prevail, although it has begun to exclude God's

---

* This can scarcely be said now. Infidelity has fearfully advanced since the time when these words were first written in 1848.

truth from governmental arrangements, excludes it, not on the ground of its being valueless, but because of the difficulty of determining what truth is: and the still greater difficulty of acting on it, if determined. Necessity therefore is made the plea. It is impossible, men say, in the present condition of society, to legislate uncompromisingly for Truth, whatever Truth may be. Expediency will not permit it. Such is men's justification of their new system; and then too it is argued, that it is a system that does practically succeed. That it should seem to succeed, is not to be wondered at; for when we remember the firm unbending character of Truth, we might expect that the course of human things would run more smoothly, when freed from its resisting influences; and smoothness is likely to be deemed prosperity, especially whilst a spurious philanthropic morality throws, for a season, a delusive brightness over the scene.

The manner in which this rising system will be developed in those very countries of the East, in which civilisation found its earliest home—its connexion with that commercial greatness which is becoming THE pillar of the nations' strength, and the recognition of this system, when established in "the Land of Shinar," by the kingdoms of the Roman world, as a federal system which they will unitedly acknowledge, are subjects on which I do not enlarge, because they are elsewhere considered.* It will be a System distinctively secular; but false religious systems, whether Jewish, or Mahomedan, or nominally Christian, (Popery, will probably be among the foremost,)† will finally bow to it, and become its handmaids. They will sell themselves to be the channels of its influence, and will thus add to their own intrinsic evil, the guilt of being the pillars of this form of concentrated ungodliness. The Harlot of the seventeenth of Revelation and "the Woman sitting in the Ephah" (Zech. v.) reveal the history of this evil system.

Antichristianism, in its full development, comes as God's scourge upon the iniquity of this system. Antichrist, when he first arises, will adopt and sustain it. The Woman is seen seated on the Beast. But as soon as he has sufficient power, as soon as the Ten Kings, having one mind, give their power and authority unto him (Rev. xvii. 13), he will turn upon the System whose servant he has been, destroy it, and become himself supreme. He will be the last centre of all governmental authority in the Roman earth, and all shall wonder after and worship him, whose names are not written in the Lamb's book of life. The liberty of the mind of

---

* See "Babylon and its Revival," and "Prospects of Ten Kingdoms," and "Europe and the East," as advertised at end.

† That is to say, Popery if it remain in the Ten Kingdoms, must *there* consent to aquiesce in the Latitudinarianism of the day. In other parts of the world it will probably maintain, or seek to maintain, its high pretensions.

man and all the other boasts of latitudinarianism will vanish then. One will have come in his own name, and God will allow him to enforce the authority of that name, and Jew and Gentile alike will bow beneath his power throughout all the appointed sphere. They will own him as their Master, and worship him as their God.

Such are the prospects of the nations in the midst of which we dwell. The tide of circumstances is setting rapidly towards this last great gathering point of evil. It is an hour when we peculiarly need watchfulness and wisdom, that we might judge every principle and every system, simply by the word of God. God's truths and God's principles, remain unchanged. He has revealed them, and revealed them plainly, by His holy Apostles and Prophets, and blessed are they who cleave to them, and seek to walk according to their testimonies. In the midst of the Antichristian darkness of the closing hour, there are some who will keep and testify to Truth. They are mentioned in the Revelation as those "who keep the commandments of God, and have the testimony of Jesus" (Rev. xii. 17); and even in heaven, their faithfulness is remembered with praise and with thanksgiving, as having "overcome because of the blood of the Lamb, and because of the word of their testimony; and they loved not their life even unto death." (Rev. xii. 11.)

CHAPTER VII.

PASSAGES OF SCRIPTURE RESPECTING ANTICHRIST COMPARED.

THE passages which have been quoted in the preceding chapter to prove the futurity of Antichrist have been chiefly taken from the Book of Daniel. But Daniel is not the only one of the Old Testament Prophets who refers to the actings of this Wicked One and his connexion with the Land of Israel. In the eleventh of Zechariah we find one remarkable passage which is, as yet, manifestly unfulfilled. After the Prophet had described the rejection of the true Shepherd of Israel, and His valuation at thirty pieces of silver, the prophecy continues thus: "And the Lord said unto me, Take unto thee yet the instruments of a foolish shepherd; For, lo, I will raise up a shepherd IN THE LAND, which shall not visit those that be cut off, neither shall seek the young one, nor heal that that is broken, nor feed that that standeth still; but he shall eat the flesh of the fat, and tear their claws in pieces. Woe to the idol shepherd that leaveth the flock! the sword shall be upon his arm, and upon his right eye: his arm shall be clean dried up, and his right eye shall be utterly darkened." These words are evidently unfulfilled, for since the Jews were scattered by the Romans no such person has been raised up to be their shepherd "*in the Land.*"

In Isaiah, also, several prophecies are found respecting him. Thus in the eleventh chapter, after describing what the Messiah of Israel shall be, and what the establishment of His kingdom will effect, when "He shall judge the poor with righteousness, and reprove with equity for the meek of the earth"—the Prophet adds, "He shall smite the earth with the rod of His mouth, and with the breath of His lips shall He slay the Wicked One." (Isaiah xi.) That these last words refer to Antichrist, is manifest from the use made of them by the Apostle in the Thessalonians.

{ And with the breath of His lips shall He slay the Wicked One.
  וּבְרוּחַ שְׂפָתָיו יָמִית רָשָׁע

{ The Wicked One whom the Lord shall consume by the breath of His mouth.
  Ὁ ἄνομος ὃν ὁ Κύριος ἀναλώσει τῷ πνεύματι τοῦ στόματος αὐτοῦ.

In the tenth chapter of Isaiah, we find a very full description of the actings of this last great scourge of Israel. The concluding verses of the ninth and the commencing verses of the tenth chapter foretell the long series of successive chastisements which the Lord would send upon the wickedness of Israel. Each clause of the description concludes with those oft-repeated words: "For all this His anger is not turned away, but His hand is stretched out still."

But after the fourth verse of the tenth chapter these words are repeated no longer, because the hour of the *last* infliction on Israel will have come. A king of Assyria is to be raised up by whom "the Lord will perform *His whole work* upon Mount Zion and on Jerusalem" (verse 12), and at the time of whose destruction "*the indignation shall cease*" (verse 25). The rise of this person, therefore, must be future; for the Lord has not yet "performed His whole work" against Jerusalem, neither has the time of the "*last indignation*" come. His commission against Israel is thus described: "I will send him against an hypocritical nation, and against the people of my wrath will I give him a charge, to take the spoil, and to take the prey, and to tread them down like the mire of the streets. Howbeit he meaneth not so, neither doth his heart think so; * * * wherefore it shall come to pass, that *when the Lord hath performed His whole work upon Mount Zion and on Jerusalem*, I will punish the fruit of the stout heart of the king of Assyria, and the glory of his high looks. For he saith, By the strength of my hand I have done it, and by my wisdom; * * * Therefore shall the Lord, the Lord of Hosts, send among His fat ones leanness; and under his glory he shall kindle a burning like the burning of a fire. And the light of Israel shall be for a fire, and his Holy One for a flame: and it shall burn and devour his thorns and his briars (*i.e.*, all his defences) in one day; .... And it shall come to pass in that day, that the remnant of Israel, and such as are escaped of the house of Jacob, shall no more again stay upon him that smote them; but shall stay upon the Lord, the Holy One of Israel, in truth. 'The remnant shall return, even the remnant of Jacob, unto the mighty God. For though thy people Israel be as the sand of the sea, yet [only] a remnant of them shall return: the consumption decreed shall overflow with righteousness. For the Lord God of Hosts shall make a consumption, even determined, in the midst of all the Land." No words can be plainer than these. They *cannot* be fulfilled until "the remnant of Jacob shall return unto the mighty God," when "the Lord, the Lord of Hosts, shall lop the bough (of Gentile power) with terror: and the high ones of stature shall be hewn down, and the haughty shall be humbled." In other words, the day of man will have ended, and the day of the Lord have come.

The last verse of the ninth of Daniel contains a virtual quotation from the passage just cited from Isaiah x. The words of Daniel, literally translated, are as follow: "And he (the prince that cometh) shall confirm a covenant with the many

SCRIPTURE RESPECTING ANTICHRIST. 119

(*i.e.*, with the multitude) for one hebdomad ; and at half the hebdomad, he shall cause sacrifice and oblation to cease, and upon the pinnacle of abominations (*i.e.*, the idolatrous pinnacle) shall be that which causeth desolation; even until the consummation, and that determined, be poured upon the causer of desolation."

The words "*consummation*" and "*that determined*," show that the two passages in Isaiah and Daniel speak of the same event, and of the same period. In the Hebrew this connexion is clearly seen :

{ כִּי כָלָה וְנֶחֱרָצָה אֲדֹנָי יְהוִה צְבָאוֹת עֹשֶׂה
For a consummation, and that determined, the
  Lord Jehovah of Hosts maketh. (Is. x. 23.)

{ וְעַד־כָּלָה וְנֶחֱרָצָה
Even until the consummation, and that determined. (Dan. ix. 27.)

"The King of Assyria," "the Prince that shall come," and "the Desolater," are thus shown to be the same.

The following quotations identify "the King" of the eighth chapter of Daniel, with "the Prince that shall come" in the ninth chapter.

{ viii. 11. He shall take away the daily sacrifice.
  ix. 27. He shall cause sacrifice and oblation
          to cease.

{ viii. 19. He shall prosper in the last end of
             the indignation.
  ix. 27. * * * till that determined is poured
          on the desolater.

Thus "the King of Assyria" (Is. x.), "the Prince that shall come" (Dan. ix.), and "the King of fierce countenance" (Dan. viii.), are identified.

The following comparison proves the connexion of the eighth with the latter part of the eleventh chapter.

{ viii. 9. He waxes great toward the pleasant land.
xi. 41. He enters into the glorious land.

{ viii. 17. At the time of the end shall be the vision.
xi. 40, 41. At the time of the end shall he enter.

{ viii. 19, 24. He prospers in the last end of the indignation.
xi. 36. He shall prosper till the indignation be accomplished.

Thus "the King of Assyria" (Is. x.), "the Prince that shall come" (Dan. ix.), "the King of fierce countenance" (Dan. viii.), and "the King who shall do according to his will," are identified.

A comparison of Dan. xi. and Dan. vii. identifies "the King who shall do according to his will" and the little horn.

{ vii. 25. He shall speak great words against the Most High.
xi. 36. He shall speak marvellous things against the God of gods.

SCRIPTURE RESPECTING ANTICHRIST. 121

{ vii. 25. He changes times and laws.
xi. 37. He shall not regard the God of his fathers, * * * nor any god, but shall magnify himself above all.

{ vii. 22. He shall continue until the saints possess the kingdom.
xi. 36. He shall prosper till the indignation be accomplished.

By a comparison of passages already made (page 101) the Little Horn of the seventh of Daniel has been shown to symbolise the same person as the ten-horned Beast of the thirteenth of Revelation. Thus "the King of Assyria" (Is. x.), "the Prince that shall come" (Dan. ix.), "the King of fierce countenance" (Dan. viii.), "the King who shall do according to his will" (Dan. xi.), the "little horn" (Dan. vii.), and the ten-horned Beast (Rev. xiii.), are shown to be identical. The following summary clearly proves that the period spoken of in these several chapters is the same:

"The King of Assyria" (Is. x.)—
    period— "When the indignation shall cease" (v. 25), and "when the Lord shall have fulfilled His whole work upon Jerusalem." (v. 12.)

"The Prince that shall come" (Dan. ix.)—
    period— "Until the consummation, and that determined be poured upon the Desolater." (v. 27.)

"The King of fierce countenance" (Dan. viii.)—
    period— "The last end of the indignation." (v. 23.)
"The King who shall do according to his will" (Dan. xi.)—
    period— "Till the indignation be accomplished." (v. 36.)
"The little horn" (Dan. vii.)—
    period— "Until the time comes for the saints to possess the kingdom." (v. 22.)
The ten-horned Beast (Rev. xiii.)—
    period— Until the gathering at Armageddon, and the coming forth of the Lord, with the armies of Heaven following.

In the fourteenth of Isaiah, Antichrist is again called the Assyrian. "The Lord of hosts hath sworn, saying, Surely as I have thought, so shall it come to pass; and as I have purposed, so shall it stand: That I will break the Assyrian *in my Land*, and upon my mountains tread him under foot: then shall his yoke depart from off them, and his burden depart from off their shoulders"—*i.e.*, the shoulders of Israel.

In the same chapter he is also called "the King of Babylon," and "Lucifer" (the Bringer of light), because he will blasphemously assume the character of Christ as the bright and morning star.*

His connexion with Mount Zion, and the Temple

---

* For the nature of his connexion with Babylon and Assyria, see "Babylon—Its Revival" as advertised at end.

of God at Jerusalem, is distinctly marked in the following passages:

Isaiah xiv. "I will sit also upon the Mount of the congregation (Zion), in the sides of the north."

Dan. viii. "By him the daily sacrifice was taken away."

Dan. ix. "Upon the pinnacle of abominations shall be that which causeth desolation."

Dan. xi. "He shall plant the tabernacles of his palace in the glorious holy mountain."

Dan. xii. "From the time that the daily sacrifice shall be taken away, and the abomination [idol] that maketh desolate set up," etc.

Matt. xxiv. "When ye shall see the abomination of desolation (desolation-idol) spoken of by Daniel the prophet, stand in the Holy Place," etc.

2 Thess. ii. "He shall sit in the Temple of God, showing himself that he is God."

The idol or abomination (for the word is the same in Hebrew) which is referred to in the preceding passages, is more fully described in Rev. xiii. It is there spoken of as the great object of worship throughout the whole extent of Antichrist's dominions—"And it was given unto him [the false

Prophet] to give breath unto the Image of the Beast, that the Image of the Beast should both speak, and cause that as many as would not worship the Image of the Beast should be killed."

It is evident that none of these things have as yet been fulfilled either at Jerusalem or elsewhere. It is true indeed that every form of influential wickedness that has yet appeared, has displayed some feature of evil that will finally be found in Antichrist; for he will be the concentration of all evil. Sennacherib, Nebuchadnezzar, Alexander, Antiochus, Herod, may all be regarded as types or forerunners of the last great Destroyer. God has appointed them to be this, in order that we might be forewarned. The same may be said of systems like Popery, and individuals like Napoleon, who have appeared since the thread of the prophetic narrative of Scripture was broken at the destruction of Jerusalem. Many of the principles of Antichrist were no doubt developed in the latter; and many of the features of the false Prophet who will minister in the presence of Antichrist are traceable in Popery. But none of these things, nor anything that has yet appeared among men, answer to the *specific* descriptions by which Scripture identifies the last great Destroyer. His reign is to be but for a little season; and with him the triumph of iniquity will end for ever.

The two following quotations sufficiently show

how distinctly the expectation of Antichrist, and his future connexion with the Jews, was retained in the early ages of Christianity; even after the progress of corruption had perverted or destroyed other branches of truth. The first is from *Cyril*, who was Bishop of Jerusalem in the fourth century. The second is from *John Damascene*, who lived in the end of the seventh and beginning of the eighth century.

CYRIL.—But this afore-mentioned Antichrist comes when the times of the sovereignty of the Romans shall be fulfilled, and the concluding events of the world draw nigh. Ten kings of the Romans arise at the same time, in different places perhaps, but reigning at the same period. But after these, the Antichrist is the eleventh, having by his magic and evil skill, violently possessed himself of the Roman power. Three of those who have reigned before him, he will subdue —the other seven, he will hold in subjection to himself. At first he assumes a character of gentleness (as if a wise and understanding person) pretending both to moderation and philanthropy, deceiving, both by lying miracles and prodigies which come from his magical deceptions, the Jews, as if he were the expected Messiah. Afterwards, he will addict himself to every kind of evil, cruelty, and excess, so as to surpass all who have been unjust and impious before him, having a bloody and relentless and pitiless mind, and full of wily devices against all, but especially against us Christians. But after having dared such things for three years and six months, he will be annihilated by the second glorious coming from heaven of the Only begotten Son of God, even Him who is our Lord and Saviour, Jesus, the true Messiah, who, having destroyed Antichrist by the Spirit of His mouth, will deliver him to the fire of Gehenna.

JOHN DAMASCENE writes as follows: We ought to know that THE Antichrist must come. It is true, indeed, that every one that doth not acknowledge that the Son of God hath come in the flesh, and that He is perfect God, and hath become (in addition to being God) perfect man, is *an* antichrist—nevertheless in a peculiar and especial manner he is called Antichrist, who is to come at the conclusion of the age. Accordingly, the Gospel must first be preached among all the Gentiles, even as the Lord hath said, and then Antichrist will come for the conviction of the rebellious Jews, for the Lord said to them: "I have come in the name of my Father, and ye receive me not—another is coming in his own name, and him ye will receive." So also the Apostle: "Because they received not the love of the truth, that they might be saved, and for this cause God shall send upon them strong delusion," etc. Accordingly, the Jews did not receive Him who is the Son of God, our Lord and God, Jesus Christ; but the Deceiver who calleth himself God, him they will receive. That he will call himself God, the angel when teaching Daniel signifies thus— "The gods of his fathers he (Antichrist) will not acknowledge," etc. So also the Apostle: "Let no one man deceive you by any means," etc. After quoting this passage from the Thessalonians, John Damascene continues, "By the Temple of God, he means not ours, but the ancient one of the Jews; for he will come not to us, but to the Jews. He will come not on behalf of Christ, but against Christ and those who are Christ's, therefore also he is called Antichrist.* (Page 299—Vol. I.)

* For further quotations on this subject see "Prospects of the Ten Kingdoms."

CHAPTER VIII.

THOUGHTS ON MATTHEW XXIV.

FROM all that has been said in the preceding chapters, it is obvious that the great leading events of "the end of this age" (Matt. xiii. 40) have been revealed very plainly in the Old Testament Prophets. They have spoken of the final visitation on Jerusalem—of the destruction of the Gentile power now paramount in the earth—of the manifestation of the Messiah of Israel in glory—and of the forgiveness and restitution of Jerusalem. Consequently, these events must have been familiar to the expectations of all who knew what God had spoken through the Prophets of Israel.

When, therefore, the disciples of the Lord Jesus stood around Him, as He was quitting for the last time the gates of Jerusalem, and heard Him speak of the desolation that was about to come upon that City and its Temple, it was most natural that their thoughts should instantly revert to the testimonies of the Prophets respecting the closing events of the age—that they should think of what Daniel and Zechariah had written, respecting the "last end of the indignation," and imagine that the Lord was speaking respecting that same final hour of visitation. How could they have thought otherwise, with the knowledge they then possessed? Hence, as soon as He had spoken of the desolation of the Temple, they instantly said, "Master, when shall these things be?"—that is, when shall this destruction, of which Thou hast spoken, fall upon this Temple?—and, "what shall be the sign of Thy coming, and of the end of the age?"

They instantly connected (and herein lay their error) the destruction of the Temple with "the end of the age," and with the manifestation of the Lord in glory. They imagined that when His prediction respecting the overthrow of the Temple should be accomplished, He would appear in His glory and terminate the indignation against Jerusalem by the destruction both of its Gentile oppressors and of His own adversaries in Israel. It was this error that the Lord Jesus, in the succeeding prophecy, sought to correct. He had no desire to lessen in their souls their apprehension of those great events, which are yet to occur at "the end of the age." He did not wish that they should forget one word which the Prophets had spoken respecting that all-important crisis in the history of Israel and of men; but He *did* desire to correct their mistake as to the *time* of His appearing, and to enlarge their knowledge respecting the circumstances that were first to come to pass.

The history of the present Dispensation — the

Dispensation in which we live—was that of which the disciples were chiefly ignorant. They imagined that the time of suffering and reproach for the servants of God was nearly concluded, and that the kingdom of God was about immediately "to APPEAR." They were uninstructed in the history of Christianity through the long eighteen hundred years which have since witnessed its rise and its corruption. The long absence of their Master from the earth—the character of the service in which they were about to be employed, in preaching the Gospel of the grace of God, and all the attendant sorrows, were as yet unknown to them. Of the corruptions of professing Christianity, and how it would combine both with Judaism and national Gentilism in giving birth to the antichristian abominations of the latter day—of all this they knew nothing. They thought that the evil generation then around them was almost immediately to pass away from before the presence of their Master, manifested as the King of Israel in His glory; and that it was to be succeeded by that new generation who should "know the Lord," whom men should call "priests of the Lord, and ministers of their God."

Accordingly, the first object of the Lord Jesus in this prophecy, was to instruct them as to the character of the period, in which they themselves were about to serve and to suffer. He knew that wars were about to come upon the land of Israel, which would result in a more entire desolation of Jerusalem than any that had yet been. He knew that the sorrows thus about to come, would in many respects resemble those more terrible calamities which are to conclude the age; and that His disciples, if not forewarned, would be likely to regard these early sorrows (especially when Satan had raised up false Christs around them) as signs of the near approach of the end. Against this He guards them. He tells them to beware of regarding the coming wars and commotions as signs of the end: "And Jesus answered and said unto them, Take heed that no man deceive you; for many shall come in my name, saying, I am Christ, and shall deceive many. And ye shall hear of wars and rumours of wars; see that ye be not troubled, for all these things must come to pass. BUT THE END IS NOT YET."

It would indeed have been joy to the disciples if these events had ushered in the end. They would have rested from their labours and received their crown of life, and been numbered among those "saints of the high places" to whom, it had been said in Daniel, "the kingdom, and dominion, and greatness of the kingdom under the whole heaven shall be given." Creation would have been freed from the bondage of corruption, and all the trees of the wood have clapped their hands. Nation would have ceased to rise up against nation, neither would they have learned war any more. The Lord would have opened His hand to satisfy the desire of every living thing. The Lord of Hosts would have reigned in Mount Zion and in Jerusalem, and before His ancients gloriously. All this would have been joy to the Disciples; but they would have missed their *distinctive* honour and blessedness. That to which they were specially called was to serve in the midst of enemies, to know suffering, trial, and reproach. They were to suffer first, and to reign afterwards.

But how different was the scene about to be opened before them. Christianity, after appearing and being rejected in Jerusalem, was (like another Naomi, driven from the land of her birth) to go down and complete a sorrowful sojourn among the Gentiles, there to lose its manhood and its strength. The nations, unreached by its gracious testimony, were to continue fierce monsters, devouring one another. "Nation," said our Lord, "shall rise against nation, and kingdom against kingdom." The relation of God to the earth, instead of being one of Millennial peace, was to be a relation of judgment. He would be obliged, because of iniquity, to send the earthquake, the pestilence, and the famine. "There shall be famines and pestilences and earthquakes in divers places." The relation of the nations to the Church was to be one of fierce and destructive persecution. "They shall deliver you up to be afflicted, and shall kill you, and ye shall be hated of all the Gentiles (τῶν ἐθνῶν) for My name's sake." And in the midst of these trials the energy of the Church itself would decay; evil would arise in its own bosom; and because iniquity should abound, the love of the greater part (τῶν πολλῶν), even of Christ's own people, would wax cold; and it would be found a hard thing to hold fast unto the end. "He that shall ENDURE to the end, the same shall be saved."

Such is the picture, drawn by our Lord Himself, of the Dispensation in which we live. And how truly has it been verified! Yet this is the picture which we, Gentile Christians, have ventured to reverse. "Wise in our own conceits," we have not only borne false witness to Israel, and hidden the prophecies which speak of the judgments about to fall on them, and of the succeeding glory, but we have perverted also, or buried out of sight, the prophecies which speak of *our own* corruptions. We have said that there has been, and shall be, both in the Churches and in the world, progress of light and increase of blessing; and neither the most obvious facts nor the plainest testimonies of Scripture have availed to awaken us from the delusive dream. We turn to the countries where St. Paul and the Apostles first spread the light of Christianity, and we behold black, settled darkness; we look upon the western nations of Europe, and we see worldliness—idolatry—infidelity; we look to the condition of real Christianity, we discern unsettlement—discord—love waxing cold; and yet many still persist in saying that progress is being made—progress according to Christ and to God; and some even assert that the Christianity of the present day stands in favourable contrast with that of the Apostolic period. But the Lord does not so judge. He has said, "As it was in the days of Noah, so shall it be when the Son of Man is revealed." (Matt. xxiv. 37.) "This know, that in the last days perilous times shall come." "Evil men and seducers shall wax worse and worse." (2 Tim. iii. 13.) Such is the picture drawn by the word of God, and verified by each day's experience. Nevertheless, in the midst of all this ruin "the Gospel shall be preached for a witness"—(observe the word, *for a witness*)—throughout all the world (ἡ οἰκουμένη), and then shall the END come.

Here the first division of this prophecy concludes. The Lord had given a brief and rapid statement of the general character of events throughout the Dispensation up to the "time of the END." It is usual in Scripture to give general descriptions first. They are given first, and extend over the whole of the period spoken of, and then the same ground, or part of it, is retraced with more particularity of detail. Accordingly, the next division of this chapter, extending from the 15th to the 28th verses inclusive, is as specific as the former part had been general. It refers to a period yet future—the period termed in Daniel "the time of the end," when Jerusalem shall again be inhabited by her own people, and the iniquity of Israel, of the Gentiles, and of a large portion of professing Christianity, be gathered round a common centre of apostasy in the worship of Antichrist in Jerusalem.

And here we arrive at that part of the prophecy in which the words of our Lord agree with those of Daniel and the Prophets of the Old Testament. He had indeed added new things; for "a scribe well instructed unto the kingdom of heaven bringeth out of his treasures things *new* and *old*." He had given the moral history of the lengthened period in which we live—a period respecting which the Old Testament affords no detail, and now, in proceeding to speak of the hour which immediately precedes His appearing, He would necessarily dwell on those events which form the great theme of the Old Testament prophecy when treating of "the time of the end."

I have already referred, in the previous chapters, to the testimonies of Isaiah, Daniel, and Zechariah, respecting that "time of the end." We have seen that they all speak of Israel, still *unconverted*, being nationally in Jerusalem at the time when the Lord appears. We have seen that they speak of the Temple as existing; of its being profaned, and the idol of the Desolater being set there; and that that period is said by Daniel to be "a time of trouble, such as never was since there was a nation, even to that same time." These are the very things of which the Lord Jesus speaks. He speaks of the idol (for "abomination" is a word used in Hebrew to denote an idol) being set in the Holy Place. He speaks of the time of trouble, such as never was. His words are, in fact, a quotation of the words of Daniel. The words in Daniel are, "a time of trouble, such as never was since there was a nation, even to that same time." (Dan. xii. 1.) The words in Matthew are, "There shall be great tribulation, such as was not since the beginning of the world to this time, no, nor ever shall be." (Ver. 22.)

It seems wonderful that any doubt should ever have arisen as to the *futurity* of all that is recorded in this part of the prophecy—that is, from verse 15 to 28 inclusive. There cannot be two *unequalled* seasons of tribulation, and therefore the reference by our Lord to the words of Daniel just quoted would be alone sufficient to prove that the period of which He speaks, is *future* still. For who doubts the futurity of the words as found in Daniel? They are there expressly connected, first, with the destruction of that wicked king, who, after "planting the tabernacles of his palace in the glorious holy mountain, comes to his end, and none shall help him" (Dan. xi. 45); secondly, with the deliverance and forgiveness of Daniel's people, Israel; thirdly, with the resurrection of the saints. It is the time when the "wise shall shine as the brightness of the firmament, and they that turn many to righteousness, as the stars for ever and ever." Unless, therefore, we can say that Antichrist *has* glorified himself in Zion and been destroyed, and that Israel *has been* forgiven, and that all the sleeping saints *have been* raised, we must admit that the time of unequalled tribulation spoken of in Daniel is yet future; and if future in Daniel, it must be future likewise when spoken of by our Lord; for, as I have already said, there cannot be two *unequalled* periods of woe.*

But again, even if the reference to the words of Daniel were not conclusive (which it is, and there-

---

* The passage in Daniel to which I refer, is as follows: "And he (Antichrist) shall plant the tabernacles of his palace between the seas in the glorious holy Mountain (Zion); yet he shall come to his end, and none shall help him.

"And AT THAT TIME shall Michael stand up, the great Prince which standeth for the children of thy people (*i.e.* Israel), *and it shall be a time of trouble such as never was since there was a nation even to that same time*, and at that time shall thy people (Israel) be delivered, every one that is found written in the book. And many of them that sleep in the dust of the earth shall awake—some to everlasting life, and some to shame and everlasting contempt." For further remarks on this passage, see "Prospects of Ten Kingdoms," p. 264.

fore no other argument *need* be added), yet a careful examination of the words of our Lord, as recorded in Matthew, would sufficiently prove the same thing. For His mention of the unequalled season of tribulation is His reply to the enquiry of His disciples respecting the sign of His coming. They had asked, "What shall be the sign of Thy coming, and of the end of the age?" In the first part of the prophecy He tells them of certain events which are *not* the sign of His coming (ver. 6), but in this second part of His prophecy, He speaks of this unequalled season of tribulation *as being* the great constituted sign. "IMMEDIATELY"—(no word can be more emphatic; it is the emphatic word of the clause)—"immediately after the tribulation of those days shall the sun be darkened, and the moon shall not give her light, * * * and then shall appear the sign of the Son of Man in heaven." And seeing that there is at this present moment, no *such* season of tribulation ("a tribulation such as never was") resting either upon Jew or Gentile, that the Jews are being relieved even of that lighter pressure which has for ages borne on them, and that they are "to be holpen with a little help, and that many will cleave to them with flatteries" (Dan. xi. 34), it follows that the unequalled season of sorrow, which is to be *immediately* followed by the appearing of the Lord, must be future.

Nor would the argument be less conclusive, if

founded simply on the words "abomination of desolation." Since our Lord spake these words, no idol has ever been set in the sanctuary at Jerusalem. The Romans, even if they had desired it (which they did not) would have been unable to set an idol there; for during the capture of the city, the Temple was, by means of a firebrand accidentally thrown by a Roman soldier, burnt to the ground. The sanctuary therefore was destroyed, and it has never since been restored; consequently, neither Roman, nor Saracen, or Turk, would have been able (even if they had wished it) to plant an idol there.*

But this argument becomes even more conclusive when connected with the words of Daniel: and it must be so connected, because our Lord Himself says, "when ye see the abomination of desolation, *spoken of by Daniel the Prophet.*" (Ver. 15.)

There are three passages in Daniel which speak of an idol in connexion with the temple of Jerusalem. A King, named Antiochus Epiphanes, who lived about a hundred and sixty-eight years *before* our Lord, placed an idol in the Temple of God at Jerusalem, and caused it to be worshipped there. This is referred to in the 31st verse of Daniel xi. where it is said that this king, and they who are with him "shall pollute the sanctuary of strength, and

---

* It is remarkable that the Romans are referred to in Daniel as "*destroying* the city and sanctuary" (chap. ix. 26), but not as planting the abomination.

shall place the abomination that maketh desolate." But seeing that this was done more than a century and a half before our Lord was born, the *prophecy* of our Lord in Matthew can have no reference thereto. It was a fact accomplished, and therefore incapable of being predicted.*

* The history of Antiochus is important, as being evidently intended to foreshadow the actings of the last great Gentile persecutor in Jerusalem. The first chapter of the first of Maccabees, which is perfectly authentic, should be read throughout. It gives a vivid impression of what the future scene in Jerusalem is to be. Antiochus is there said to have "set up the abomination of desolation," and "sacrifice was offered upon the idol altar which was upon the altar of God."

"Now the fifteenth day of the month Casleu, in the hundred and forty and fifth year, they set up the abomination of desolation upon the altar, and builded idol altars throughout the cities of Judah on every side—and burnt incense at the doors of their houses and in the streets. And when they had rent in pieces the books of the law which they had found, they burnt them with fire. And wheresoever was found with any the book of the Testament, or if any consented to the law, the king's commandment was that they should put him to death. Thus did they by their authority to the Israelites every month, to as many as were found in the cities. Now the five and twentieth day of the month they did sacrifice upon the idol altar which was upon the altar of God. At which time, according to the commandment, they put to death certain women that had caused their children to be circumcised; and they hanged their infants about their necks, and rifled their houses and slew them that had circumcised them." I Maccabees i.

These things had occurred not very long before the Lord

The other two passages in Daniel, in which the "abomination of desolation" is mentioned, refer to a time which is still future. The first of these is Daniel ix. 27, of which the literal translation is as follows: "And upon the pinnacle of abominations (*i.e.* the idolatrous pinnacle) shall be the Desolater, even until the consummation and that determined be poured upon the Desolater." In the preceding paper, I have remarked on this passage, and on the force of the words "consummation" and "that determined."* Unless therefore we can prove that the close of Jewish desolation has been already reached, and that all "that is determined" has been poured upon the last great "Desolater," it follows that the setting up of the abomination of desolation, is an event yet to be accomplished.

The other place in which the abomination of desolation is mentioned, is in the 11th verse of chapter xii. This must be future, because the *completed* period of Jewish blessing is mentioned as occurring 1335 days after it has been set up. Since therefore the words of our Lord in Matthew cannot refer to Antiochus, who lived 160 years before

---

Jesus was born. They were fresh in the remembrance of Israel. They well understood what "abomination of desolation" meant. When the Lord therefore used that expression, it was one which the disciples found little difficulty in apprehending.

* See p. 119.

the Lord Jesus came, they must be quoted from one of the other passages, and both these are future, and belong to the close of Jewish desolation.

And can we, after all that we have read in Isaiah and in Daniel respecting the history of Antichrist and of Israel in the latter day, doubt respecting the meaning of the passage before us? Are not the Jews to return to Jerusalem in unbelief? Will they not rebuild their Temple? Are they not quite ready to welcome any one who shall come to them, glorious in this world's glory, and to make him their staff whereon to lean? They would *not* receive Jesus of Nazareth, their King. They saw no beauty in Him that they should desire Him; but when another shall come to them (as this great messenger of Satan will come) in his own name, him they will receive. (John v. 43.) We read in the Revelation of his glory, for he is to wear the ten diadems of the Roman world—we read also of his Image, and of its being worshipped. How then can we doubt that the abomination of desolation will indeed stand in the holy place, and that Antichrist will be worshipped there. " He as God shall sit in the Temple of God, showing himself that he is God." (2 Thess. ii. 4.)

It is possible that men now living may behold these things; for the time is at hand. But however this may be, there will be some " holding the faith of Jesus," who will be in Jerusalem at that hour. They may tarry there until then, but as soon as this consummation of blasphemy takes place, they are commanded instantly to depart. They are commanded *instantly* to flee, because the Lord God of Israel—He who will have been so long outraged and contemned, will at last pour forth the *fulness* of His wrath upon that Land and City, and will send upon Jerusalem tribulation such as never was since there was a nation, even to that same hour." His chastening hand will deal in overwhelming fury with Israel then: "You only have I known of all the families of the earth: therefore I will punish you for all your iniquities." (Amos iii. 2.) "Except those days should be shortened, no flesh should be saved, but for the elect's sake those days shall be shortened." (Ver. 22.)

The merciful loving-kindness of the Lord Jesus desires that His people, even all who believe in His name, should be withdrawn from the sphere on which the weight of this sore tribulation is to fall. He directs them, therefore, to quit Judæa and Jerusalem: and, as if to show the minuteness of His care, He directs them to pray that things may be so ordered, as that neither local circumstances, nor the inclemency of winter might impede their way. "Pray ye that your flight be not in the winter, neither on the Sabbath day." (Ver. 20.) He will not cease to number the hairs of the head of His people: themselves, their families, their children will still be the subjects of His solicitude and care. At a past period,* when the days of vengeance first began to close in upon Jerusalem, the servants of Christ were commanded to flee (see Luke xxi. 21), and they did flee, and a Pella was provided for them. So will it be again in the yet deeper season of coming woe. The circumstances may be different, the tribulation more intense, faith perhaps weaker: but there will be the same everlasting arm of power, the same heart of faithfulness, that shall protect in future even as in past sorrows.

"IMMEDIATELY after the tribulation of those days shall the sun be darkened, and the moon

---

* The twentieth verse of Luke xxi., which has commonly been confounded with the fifteenth of Matthew xxiv., has no connexion with it whatsoever. More than eighteen centuries separate the two verses. Luke speaks of "Jerusalem being compassed with armies," *i.e.* the Roman armies under Titus, and then details the sufferings and the scattering of the Jews which have since followed. Matthew speaks of no armies, but of the abomination being set up by one who is in possession of the city in which he quietly perpetrates this evil. The early Christians, in obedience to the command in Luke, did quit Jerusalem when they saw the Roman armies collecting against it; but it would be too late to flee from armies when they had already captured the city, and were in possession of it, and of the Temple. Accordingly, in Matthew, the reason of the flight is the coming judgments of God—not the siege of armies. Luke (xxi.) in the 20, 21, 23 and 24 verses speaks of the past; Matthew xxiv. from verse 15 and onward, of the future tribulation that is to come on Jerusalem.

shall not give her light, and the stars shall fall from heaven, and the powers of the heavens shall be shaken. And then shall appear the sign of the Son of Man in heaven: and then shall all the tribes of the Land (πᾶσαι αἱ φυλαὶ τῆς γῆς) mourn, and they shall see the Son of Man coming in the clouds of heaven, with power and great glory. And He shall send His angels with a great sound of a trumpet; and they shall gather together His elect from the four winds, from one end of heaven to the other." (Ver. 29—32.)

These words require little comment. They teach, what indeed all Scripture teaches, that Israel, as a whole, will be unprepared for His coming. "All the tribes of the land shall wail at Him. Even so, Amen." (Rev. i. 7.) The signs in the heavens, the darkening of the sun, and moon, and stars, the shaking of heaven and earth, are always mentioned as the accompaniments of the hour of His appearing. "But now He hath promised, saying, Yet once more and I shake not the earth only, but also heaven." (Heb. xii. 26.) I may here refer to the remarks I have already made on the fourteenth of Zechariah in confirmation of these things.

The age was not to end—the generation then present was not to pass away, until all these things had first been fulfilled. The individuals indeed, then living and acting, were to die. Herod and Pilate, Annas and Caiaphas, were to sink into the grave—but they were to be succeeded by others, like themselves, who would still maintain the same moral character of the "generation," and make it, in the estimate of God, one unchanged *evil* generation, even to the end. Neither in the ordinary arrangements of men, nor in the Scripture, are corporate bodies supposed to die. They may be abolished; but as long as they are permitted to remain, their existence and action corporately, is entirely irrespective of the presence or absence of the individuals of whom they may have been originally composed. They have a corporate identity, unaffected by mere circumstantial variations. The generation, therefore, that rejected the Lord, is still considered to exist; and it will not pass away and be succeeded by that new generation, which shall know the Lord, "from the least of them unto the greatest of them" (Jer. xxxi. 34), until the things of which the Lord has here spoken have been fulfilled.*

It is not my desire at present, to dwell on the

* The contrast, between the generation which is to continue till the end of the age, and the *new* millennial generation, is sustained throughout the Scripture. The thought of "newness" is so marked, that in the Psalms the millennial generation is even called "a people that shall be created." "This shall be written for the generation to come, and the people that shall be created shall praise the Lord." (Psalm cii. 18.) Chapter lxv. of Isaiah is almost equally distinct. "And ye [that is, you who compose the old generation] shall leave your name for a curse unto my chosen;

practical consequences that may flow from these things—my object chiefly is, to awaken attention to the facts of this prophecy, especially to the great future facts which mark the character of this Dispensation at its close. It can scarcely surprise any who mark what is now passing in the world, to be told, that the present convulsions will at last terminate in establishing a tenfold division of territory among the countries once reigned over by Rome; that similarity of institutions and laws will prevail throughout the countries thus divided, and that infidelity will reign in their midst. He must be blind indeed, who discerns not already its rising

---

for the Lord God shall slay thee, and call His servants by another name."

The same corporate sense attaches to words in other parts of this prophecy. When our Lord quitted the gates of Jerusalem, and saw certain persons before Him who stood as the representatives of Israel nationally, He said, " *Ye* shall not see me till ye shall say, Blessed," etc. They have not yet said this; so that "ye," in its *corporate* sense, has already extended over more than 1800 years. So also it was said to the Apostles, who were the representatives of Christians, "When ye shall see the abomination of desolation." Personally, they never lived to see it, but those whom they represented will. So also St. Paul says, " *We* which are alive and remain unto the coming of the Lord." He *knew* that he was to die. It is the corporate "*we*."

For further remarks on the force of the word "generation," see " Prophecy of the Lord Jesus in Matthew xxiv. and xxv. considered" (pages 73, 74, etc.), as advertised at end.

floods. It is difficult to understand how any one, judging even from present appearances, can doubt that iniquity will prevail over the most civilised and influential nations of the earth; and that of them it may be peculiarly said, "as it was in the days of Noah, so shall it be when the Son of Man is revealed."

But whilst such is, and will be, the condition of the nations, there is a relation to Christ that may be held in the midst of them, which can be owned by Him, even in these days of weakness, as according with His will. "Who then is that faithful and wise servant, whom his Lord shall make ruler over his household, to give them their portion of meat in due season?" There may be households— gatherings of the family of faith, having over them wise and faithful servants, dispensing to each his portion of meat in due season. With such may we desire to be. Whether we be among the teachers or the taught, it is the position that the Lord has pronounced blessed. Association in the name of the Lord Jesus (*when its principles are not self-devised and arbitrary, but the principles of His revealed will,*) stands in such contrast with the confederated lawlessness of the latter day, that it cannot be valueless in the sight of Him who does not "despise the day of small things."* That the general condition of real

* Simply to gather together the children of God is one thing : *so* to gather them together that they should be ser-

Christianity throughout the whole earth, will be one of slumber at the close, is too plainly taught in the parable of the virgins to admit of question. But because Christianity, as a whole, slumbers, therefore we, as individuals, are exhorted to watch and to trade with our talents, not believing that our Master is austere or hard, gathering where He has not strawed, but that His grace is such, that the least act done to the feeblest of His believing people for His name's sake, shall be owned by Him, even when He shall sit upon the throne of His glory, as if done to Himself. It will stimulate us to trade on with our talent, when we remember that we are serving One who so estimates and so acknowledges what any thing but grace would scorn. If the close of this Dispensation settles down into a night of deep darkness, the more precious is any light that even glimmers in the midst of the gloom.

---

vants of and witnesses to the Truth, is quite another thing. In this last case we may expect the blessing of God. "Sanctify them through Thy truth; Thy word is truth."

CHAPTER IX.

ON LUKE XXI.

IT may seem scarcely needful, after reading the enlarged and specific prophecy in the twenty-fourth of Matthew, to dwell on the more brief and general statements in the gospel of Luke. The prophecy in Luke has been commonly considered to be a mere repetition, in a more general form, of the instructions in Matthew; and attempts have frequently been made to represent them as strictly parallel throughout.

But there are seldom, if ever, mere re-statements in Scripture. It is true, indeed, that when two different passages record the same facts, or deliver like precepts, there must be a substantial similarity between them. But even then, the addition of new circumstances, or an altered arrangement in the order of instruction, or even a change in the form of expression, may mark a character of difference important to be recognised in a careful exposition of the word of God. Passages strictly parallel are seldom, if ever, found.

So is it in the chapter before us. There is, of

course, a substantial agreement with the prophecy in Matthew. The events which were immediately to follow our Lord's departure from the earth, and also the signs—"signs in the sun, and moon, and stars," which are to usher in His return, are spoken of almost in the same language both in Matthew and Luke. But in treating of the intervening period, there is a marked difference. The great event recorded in Luke respecting Jerusalem, is, its destruction by the Roman armies. He speaks also of the events connected with that destruction, in their relation both to the disciples and to the Jews. But in Matthew, all this is so slightly touched on, that it may be almost said to be passed over in silence.* In Luke, on the other

* The only reference to this event in the prophetic part of Matthew xxiv. is in the sixth verse, where the Lord speaks of those wars which were to follow His departure from the earth, but which were NOT signs of "the end of the age."

It is quite in consistency with the character of Matthew's gospel to dwell on the last Antichristian abominations in Jerusalem, and pass over in comparative silence its past destruction: for the gospel of Matthew is specially concerned with the history of Christianity and its corruptions. After stating the rejection of the Lord by Israel and the setting aside of Israel corporately (see close of chapter xii.), it speaks in a number of prophetic parables (see chapters xiii. and xxv. also chapter xxii.) of the history of Christianity and its corporate failure. And inasmuch as the past destruction of Jerusalem was *not* caused by Christianity or its corruptions, we should not expect that much would be said respecting it;

hand, the great event which Matthew records respecting the latter-day history of Jerusalem, is not mentioned. Nothing is said respecting the "abomination of desolation," and merely a general reference made to "the Times of the Gentiles" being concluded by distress, terror, and signs from heaven. The two prophecies therefore are not identical. It is by reading them separately, and then uniting their instructions, that we learn respecting the past and coming chastisement of Jerusalem: and with the past and future history of that city, so many threads of Jewish history, and of Gentile national history, and of Gentile-Christian history are blended, that we might almost say that to know the two periods of which Matthew and Luke treat, is to know the Dispensation in which we live, as to its most important and characteristic features, from its beginning to its close.

The prophetic part of Chapter xxi. of Luke commences with the eighth verse. The eighth and ninth verses form the first division of the chapter.

---

but seeing that Antichristianism is to spring out of the very bosom of professing Christianity, and absorb into itself the iniquities of Gentiles and Jews in Jerusalem as a centre, we might expect that an event so intimately connected with the history of the professing Church, would be very distinctly referred to in Matthew.

Besides this, it may be given as a *general* rule, that every thing "*corporate*" in character is found in Matthew—that which is individual in Luke.

They are as follow: "*Take heed that ye be not deceived: for many shall come in My name, saying, I am Christ; and the time draweth near: go ye not therefore after them. But when ye shall hear of wars and commotions, be not terrified: for these things must come to pass; but the end is not immediate.*" (ουκ ευθεως.) These verses are closely parallel to the fourth, fifth, and sixth verses of Matthew xxiv., and speak of events which immediately succeeded the Lord's departure from the earth, events which were *not* to be mistaken for signs of the end. "The end is not immediate." Accordingly, the troubles in Judæa came; Jerusalem was taken, its Temple destroyed, its people scattered, and eighteen hundred years have since passed, but the end is not yet come.

The tenth and eleventh verses form the next division of the chapter. "*Then said He unto them, Nation shall rise against nation, and kingdom against kingdom: and great earthquakes shall be in divers places, and famines, and pestilences; and fearful sights and great signs shall there be from heaven.*" These verses describe the governmental relation of God to the earth during all the long period of Jerusalem's punishment, until "the Times of the Gentiles" shall have been fulfilled. Instead of its being a millennial relation of peace and blessing, the earth throughout the whole of this lengthened period is to be visited, again and again, with wars, pestilences, and famines, to be followed, at the close, by fearful sights and signs from heaven.

These fearful sights and signs from heaven have never yet been. As yet, though the earth has been the scene of many a convulsion, the heavens have remained in their tranquillity. "The sun has not yet been turned into blackness, nor the moon into blood." The still peacefulness of the firmament, as the stars nightly resume their course, has often stood in strange contrast to the unquiet restlessness of the scene below. But an hour is coming when the heavens also shall be shaken. "Yet once more I shake not the earth only, but also heaven," and then the end shall come.*

* These signs *from heaven* are again referred to in the concluding part of this prophecy (verse 25), where our Lord, after re-tracing the same period as is gone over in verses 10, 11, and re-tracing it with specific reference to the capture of Jerusalem and the dispersion of the Jews, again speaks of "the time of the end," and refers to these signs in the heavens, saying, "there shall be signs in the sun, and in the moon, and in the stars."

Some have imagined that signs of this kind were seen at the past siege of Jerusalem, and have referred to Josephus in verification of their statement. But in the first place our Lord does not, either in this or in the corresponding passage in Matthew, speak of the commencement, but of the close of the "days of vengeance." He is expressly speaking of the period when "the Times of the Gentiles" are *fulfilled:* the same period that is called in Daniel "the last end of the indignation."

Moreover, all parts of Scripture which speak of these

Having thus predicted the woes that were about to be sent upon Judæa, and also the more general

signs in the heavens speak of the period as one of universal dismay and terror over the face of the whole earth; "men's hearts failing them for fear," etc. It was anything rather than this, throughout the world generally, when the Romans took Jerusalem.

Nor does Josephus himself record anything that is worthy of being considered "signs in the sun, and moon, and stars." The pretended prodigies, recorded apparently on the report of others, are as follow :—

First, a star resembling a sword, which is said to have stood over the city, and a comet which lasted a year.

Secondly, a light, bright as day, shone around the altar and holy place, at the ninth hour of the night, for half an hour.

Thirdly, a heifer, as she was led by the High Priest to be sacrificed, brought forth a lamb in the midst of the temple.

Fourthly, the brazen gate of the inner court of the temple, when firmly fastened, opened of its own accord, and could not be shut again without great difficulty.

Fifthly, on one occasion, before sunset, chariots and troops of soldiers in their armour were seen running about among the clouds and surrounding cities.

Lastly, the priests, when going by night into the inner court of the temple, said they felt a quaking, and heard a great noise, and afterward a sound as of a multitude, saying, "Let us remove hence."

Such are the statements of Josephus, not one of which, it should be remembered, is confirmed by Tacitus, who gives an account of the siege, nor by any other historian. The habit of the Romans apparently was, to invent stories of the kind to terrify the besieged into the belief that they were

chastisements upon the nations, which extend throughout the Dispensation, the prophecy, in its next division, from the twelfth to the nineteenth verse inclusive, proceeds to speak of the early condition of the Church. "*But before all these,* [*i.e.* before the wars that are to desolate Jerusalem shall commence] *they shall lay their hands upon you, and persecute you, delivering you up to the synagogues, and into prisons, being brought before kings and rulers for My name's sake. Settle it therefore in your hearts, not to meditate before what ye shall answer: for I will give you a mouth and wisdom, which all your adversaries shall not be able to gainsay nor resist. And ye shall be betrayed both by parents, and brethren, and kinsfolk, and friends; and some of you shall they cause to be put to death. And ye shall be hated of all men for My name's sake. But there shall not a hair of your head perish. In your patience possess ye your souls.*"

These words describe the early condition of Chris-

---

forsaken by their gods. (See Livy, *passim.*) But whatever may be thought respecting these statements, it is manifest that they cannot be fulfilments either of Matthew or of Luke. Matthew says, "Immediately after the tribulation of those days shall the sun be darkened and the moon shall not give her light, and the stars shall fall from heaven, and the powers of the heavens shall be shaken." Luke says, "There shall be signs in the sun, and in the moon, and in the stars; and upon the earth distress of *nations,* [not of 'ews merely,] with perplexity, the sea and the waves roaring."

tianity, when the Church, in Pentecostal power, first took its place as God's witness in the earth. It witnessed of truth and grace. It testified remission of sins through the blood of the Cross, amongst those on whom chiefly rested the guilt of the rejection and death of the Lord Jesus. "Men and brethren, through this Man is preached unto you the forgiveness of sins: and through Him, all who believe are justified from all things." Such was the character of their testimony, even among those who had been His betrayers and murderers. Before any of the threatened chastisements were sent, either on the land of Israel or on the nations, this message was first to be given. It *was* given, and it was rejected. The governors of Jerusalem and of the nations, heard and scorned it, and at length began to persecute and destroy those who bore it. James, Stephen, and others, sealed their testimony with their blood, and then in well-deserved retribution, the days of vengeance began to settle in upon Jerusalem, and nothing but turbulence and sorrow has prevailed throughout the earth ever since. Nor can it be otherwise, for the destinies of the Gentile nations, *as nations*, are bound up with the destinies of Jerusalem. "The Deliverer must come out of Zion, and turn away ungodliness from Jacob," and then [not before] "His ways shall be known upon earth, and His saving health among all nations."

We do not find in these words of Luke any intimation, as in Matthew, of the corruption and failure of Christianity. We do not find it said, that "iniquity shall abound," or "love wax cold." At the time when the "abomination of desolation" shall be set up, it will be so. Indeed it is so already. But before Jerusalem was compassed by the Roman armies, it was *not* generally so. The Church (though it had begun to decline) had not yet lapsed and forfeited (as soon afterwards it did) its corporate standing. The Gentile Churches were still worthy of being symbolised by "candlesticks of gold." The servants of Christ were still preserving their separateness; and still remembered the words of their Lord, that, although hated of all men and given over to death, not a hair of their heads should finally perish. They knew that their hope was in resurrection, and in patience possessed their souls. This *past* description of the Church, as found in Luke, differs greatly from the description which Matthew gives of the corruption and failure of Christianity at the closing period of its history—the period on which *we* are now about to enter.

The "days of vengeance" began to settle in upon Jerusalem, as soon as the Roman armies were gathered together against it. These days of vengeance are continuing still, and will continue, until all that has been written against Jerusalem and her people shall have been fulfilled: "*When ye shall see Jerusalem compassed with armies, then know that the desolation thereof is nigh. Then let them which are in Judæa flee to the mountains; and let them which are in the midst of it depart out; and let not them that are in the countries enter thereinto. For these be the days of vengeance, that all things which are written may be fulfilled. But woe unto them that are with child, and to them that give suck in those days! for there shall be great distress in the land, and wrath upon this people. And they shall fall by the edge of the sword, and shall be led away captive into all nations: and Jerusalem shall be trodden down of the Gentiles, until the Times of the Gentiles be fulfilled."*

The disciples obeyed these directions. They saw Jerusalem compassed with armies, and they fled. A refuge was provided for them in a little city named Pella, where they were preserved. Their history remains as a memorial of the faithfulness and care of God, and will, no doubt, comfort and encourage those, who, at an hour yet future, will again be called on to quit Jerusalem, when a tribulation far greater and more terrible shall fall upon it—a tribulation not caused by human agency merely (though that will not be wanting), but by the immediate infliction also of terrors and judgments from God's own outstretched arm. "Woe is me: for that day is great: it is even the time of Jacob's trouble; but he shall be saved out of it." So "the Times of the Gentiles" shall end. Not more than twelve hundred and sixty days will pass between the setting "the abomination of desolation" in the Holy Place, and the conclusion of "the Times of the Gentiles," when the day of visitation shall come.

The conclusion of "the Times of the Gentiles" will be indicated by "*signs in the sun, and in the moon, and in the stars; and upon the earth distress of nations with perplexity at the roar of the sea and of the surge;** men's hearts failing them for fear, and for looking after those things that are coming on the world* (τῃ οἰκουμένῃ); *for the powers of the heavens shall be shaken.*" These words plainly teach us that it is these signs in the Heaven above, and in the earth beneath, that will break in upon the world's slumber, and cause all to realise that the Day of Visitation is at last come. These "signs" will harm no one. They are not in themselves judgments: they are only *signs* of judgments about to be. As soon as the "signs" appear, the cry will go forth throughout the whole of the slumbering Church, "Behold, the Bridegroom cometh." They also who dwell in the uttermost parts of the earth, in the dark heathen lands, will be "afraid" at the "tokens," though they will not at that time understand what those tokens indicate. The true people of Christ are commanded, the moment these signs appear, to

* The corrected reading is, ἐπὶ τῆς γῆς συνοχὴ ἐθνῶν ἐν ἀπορίᾳ ἤχους θαλάσσης καὶ σάλου. See Tregelles in locum.

"look up and lift up their head" in confidence and joy. "Now, when these things *begin to come to pass*, then look up, and lift up your heads, for your redemption draweth nigh." Such are the words of the Lord Jesus; and can any words be more plain? The saints, therefore, will certainly be on the earth when these signs appear; but when the Lord *comes to the earth*, they will not be there; for they are to meet Him in the air, into which He first descends, and thence are to come with Him.* Other men, indeed, "will see Him coming to the earth in power and great glory;" but the saints will have been changed and caught up to meet Him, and will have fallen into the train of His glory whilst He is yet in the air, and before He is seen *coming to the earth.* They will be with Him before one of the judgments which He comes to inflict shall be administered; consequently, they will escape all those things that shall come to pass; that is to say, they shall escape all the judgments of the Day of the Lord.

* The word ἀπάντησις, which is used in I Thess. iv. 17, denotes a meeting and coming with the person met. The word ἔπειτα, which is used in the first clause of this verse, should be translated not "*then*," but "*afterward.*" The saints are to be "changed in the twinkling of an eye" (see I Cor. xv.); but they will not be raptured or caught up in the twinkling of an eye. A *brief* interval is to elapse between their change and their rapture; in which interval the separation spoken of in Matt. xiii. 49 will take place.

They shall see the signs that usher in that day, but the signs, as I have already said, harm no one. They are tokens merely; and when they see these tokens they are commanded to rejoice: "When ye shall see these things come to pass, then look up, and lift up your heads, for your redemption draweth nigh."

Yet this, it must be remembered, cannot be said to all who name the name of Christ. In the professing Church there are some who have no oil in their lamps; some who, because of their profession, are regarded as branches in the true vine, but fruitless. The period of our Lord's return will be a time when (for certain special reasons that could easily be pointed out) multitudes in Christendom, and throughout the world generally, will be *peculiarly* absorbed in "the cares of this life," for there will be especial reasons for anxiety and carefulness then, because of certain disastrous events that will have occurred in the earth, and shattered the very pillar of its prosperity. The nations of the Roman World will have received a blow such as they never had received before; and on all who are absorbed by the anxieties consequent on that blow, that Day will come " unawares as a snare," and all those on whom it comes "as a snare" will perish.*

---

* The words, "*as a snare*," should be appended to the 34th verse; "and so that day come upon you suddenly as a snare." The next verse is, "For it [*i.e.* the Day of visita-

Therefore, we are commanded to watch and to pray always, that we may be strong and prevail against all the influences of that evil hour, and so be numbered among "the overcomers." "He that overcometh shall not be hurt of the second death." They that overcome shall escape every judgment, and shall stand before the Son of Man. They shall not "be ashamed away from Him [ἀπ' αὐτου, see I John ii. 18] at His coming." The inhabitants of the earth will, at the period of

---

tion] shall come upon all them that dwell on the face of the whole earth." (See revised Greek Test. Tregelles.) The day itself shall come upon *all men* throughout the whole earth. It shall be a day of universal visitation: but it shall not come "*as a snare,*" *i.e.* destructively on the saints; nor indeed on any of those who shall be spared and subsequently converted, as many will be, both among the Heathen and the Jews.

In the following verse instead of "that ye may be counted worthy [καταξιωθῆτε] to escape," read, "that ye may be strong [κατισχύσητε] to escape." The preposition κατὰ may be taken as intensifying the force of ισχυω, *to be strong;* or it may express the meaning "to be strong *against*" something; and might then be rendered "to prevail." In that case, however, the object against which the strength is put forth is commonly expressed. Whether we translate, "that ye may be strong to escape," or "that ye may prevail to escape," the general sense is the same. We shall need that strength which grace only can give, to enable us to prevail against the many influences that will then bear on us either to drive us, or to lure us from "abiding in Christ." They that do not abide in Him will not escape the things that will come to pass.

our Lord's return, be distributed into five great classes.

I. There will be the uncivilized Heathen; those who have not heard the fame or seen the glory of God. (See Isaiah lxvi. 19.) Of these, a large portion will be spared, and receive the Gospel of the grace of God, preached by converted Israel.

II. There will probably be multitudes of Mahommedans, Buddhists, and the like (especially in India and in Central and North-Eastern Asia), who will not be absorbed into Antichristianism, but will continue in their present forms of Apostasy. All such, seeing that they will have had the opportunity of becoming acquainted with the Bible and with the doctrines of Christ, will be under greater responsibilities than the uncivilized Heathen.

III. There will be the Jews; some in their Land, others scattered throughout the earth. Of those in the Land, "a third part" will be spared (see Zech. xiii. 8), and among those that are scattered, there will also be a remnant. These will be converted, and made a blessing in the earth; but the great majority of Israel, especially those in the Land, will have linked themselves with Antichrist, and will share his doom.

IV. There will be the Antichristian nations; that is, all the nations of the Roman World ($\tau\eta s$ $o\grave{\iota}\kappa o\upsilon\mu\epsilon\nu\eta s$), then divided into Ten Kingdoms. These, nationally and governmentally, will have renounced all recognition of God and of Christ, and (with the exception of "Edom, Moab, and the chief of the children of Ammon," see Dan. xi. 41) will be under the full control of Antichrist. These countries, however—that is to say, Moab, Edom, and part of Ammon, and also the Arabian tribes (Ishmaelites and Haggarenes, see Ps. lxxxiii. 6)—will, at the very end, associate themselves with Antichrist with the avowed object of utterly destroying Israel. "Come, and let us cut them off from being a nation; that the name of Israel may be no more in remembrance." Joining the mighty hosts of Antichrist (which will have been assembled at Armageddon for another object), they will thence unitedly advance into the valley of Jehoshaphat, and will there be trodden in the wine-press of the fury and wrath of Almighty God. (See Joel iii. 12, Zech. xiv. 12, and Rev. xix. 15.)

V. There will be Christendom—the "wheat and tare field." Multitudes throughout the earth will still continue to profess the name of Christ. Abstract the Ten Kingdoms of the Roman World, and (with that exception) Christendom will, in all probability, at that time be just what it at present is. All who shall continue (whether nominally or in sincerity) to profess the name of Christ, will be regarded by Him as included in His kingdom.

Heathendom, the Jews, and the Antichristian nations will not be dealt with in the *same* way. They will all be dealt with differently and distinctly.

But Christendom will not escape judgment, though none will fall on the saints who form a part thereof. The end of Christendom will have come. The first act of the Lord Jesus when He descends into the air will be to gather (not out of the earth generally, but out of Christendom) "all things that offend, and them that do iniquity." Speaking of the end of Christendom, the Lord Jesus says, "Thus shall it be at the end of the age; the angels shall come forth and sever the wicked out of the midst of the righteous (ἀφοριοῦσι τοὺς πονηροὺς ἐκ μέσου τῶν δικαίων), and shall cast them into the furnace of fire; there shall be weeping and gnashing of teeth." Have we not reason, then, to watch and pray that we may have strength to resist all seduction, and all the power and wiles of the Enemy, so that we might abide in Christ, and not be "ashamed away from Him at His coming"?

It is well that we should endeavour to realize the circumstances in which Christendom will find itself in the day of its visitation. Let us imagine a family of ten persons assembled in one room— five of them *real*, five *nominal* believers. Suddenly there will be signs in the sun, and in the moon, and in the stars. Immediately after, the Lord will descend into the air, and then, in the twinkling of an eye, the five who are true believers will be changed into His glorious likeness. In them all mere natural feelings will cease to be: nothing can pain them: nothing affright. Mortality will be swallowed up of life. Having been through grace made strong so as to prevail against all obstructions to an abiding in Christ, they will escape everything that has the character of judgment, whether directed against Christendom, or the Jews, or the Heathen, or the Antichristians. At the same moment that the saints are changed, holy angels will enter that room, and sever out of the midst of the changed those that remain unchanged. There will be glory and deliverance to the one: judgment to the other.

These truths have not been realized by multitudes who have spoken and written much respecting the coming of the Lord. Their attention has been so fixed on Antichrist, and on the chastisements God will send on Israel in connexion with him, and on the persecution that will fall on all the true servants of Christ who come within the scope of his power, that they have become absorbingly (might I not say, exclusively?) occupied with the question whether they themselves will, or will not, be exempted from these sorrows. As a consequence, their minds have become totally unfitted to receive the instructions which the Lord has given in the Scripture respecting the accompaniments and characteristics of the Day of Visitation. As regards the persecutions of Antichrist, they will be restricted to a comparatively small portion of the earth, for the countries over which he will rule are little more than the borders of the Mediterranean Sea; and although it is true that other countries, attracted by the blaze of his glory, may voluntarily associate themselves with his ways, and help forward his persecutions, yet God has on past occasions often shown that He has many means of neutralizing or modifying the fiercest efforts of persecuting power. And what will these persecutions be to them on whom they fall? Will they be curses? Will they be judgments from the hand of God? Will they not rather be for honour, joy, praise, and glory? Who more honoured, who more blessed, than they who shall be counted worthy to stand, faithful and true, in the last great conflict between truth and falsehood, between God and Satan? Of them it is written, "They overcame him because of the blood of the Lamb, and because of the Word of their testimony, and they loved not their lives unto the death." Such will be the thanksgiving of Heaven over their faithfulness. And, although sore judgments, the like to which never yet have been, will fall, measurably on all the Antichristian nations, and fully on the Land of Israel, yet those judgments will not be directed against the people of Christ who may be dwelling in those nations. The plagues sent on Pharaoh and Egypt were not directed against the children of Israel, whom God came to deliver. There was light in Israel's dwellings when there was darkness in the whole land besides. It is the heritage of God's people to escape destroying judgment; but to escape suffering for Christ's sake and the Gospel's, is not their heritage. To meet and to endure such suffering, is their honour and their joy.

If Luther, and they that were with him, in their conflict with the Papal Antichrist, instead of giving themselves to the ordering of those truths whereby Romanism and its lies were to be confronted, had thought merely of succour or deliverance for themselves, should we ever have been blessed with the light of Protestantism? And are there no signs of a coming conflict in the future, both with old enemies and with new? A mighty system of infidel Latitudinarianism is rising up amongst us, ready to shelter Romanism, Mahommedanism, and everything else that exalts itself against the oneness of Revealed Truth; and that system will be organised, established, and at last sustained and worked, not by *an* Antichrist, but by THE Antichrist. All who come within the scope of this system (they only excepted whose names are written in the Lamb's book of life) will give themselves over to its harlotries and its seductions. Will God, when that system is developed, raise up no testimony against it? Will He be silent? No. He will raise up a testimony more full, more pure, more blessed, than any that has been heard on earth since the Apostles died. Shall we not then bethink ourselves where and what that testimony is to be, and order ourselves and our ways accordingly? Or shall we give way to the delusion of the hour, and speak of being taken out of the earth before the hour of Antichristianism comes, and so direct our thoughts and the thoughts of others away from those very facts and truths, the knowledge of which is specially intended to strengthen the true people of Christ for their final hour of conflict? The saints have not only to do with God and the works of God: they have also to do with Satan and the works of Satan. They have to be *with* the one and *against* the other. The same book that tells us of the Heavenly City—the Bride of the Lamb—tells us also of another city that is the mother of all the harlotry and abominations of the earth. Are we to close our eyes to the characteristic course and history of that evil city?

The heart of the believer needs to be chastened by the knowledge of evil as well as comforted by the knowledge of good. To be ready for the coming kingdom of our God and Saviour in the sense of having a *valid title* to inheritance therein, can be affirmed by all who have truly rested on the finished work of the One great Substitute: but to be *practically ready* is another thing. Practical readiness can only be found in us when our thoughts, feelings, and ways are found to be right, wise, and true when tested by the Word of God.

CHAPTER X.

REMARKS ON THE PROPHETIC STATEMENTS OF MR. FLEMING.*

UPWARDS of twenty years have now passed since circumstances, into the special character of which it is not necessary to enter, awakened among many, both in this country and abroad, an anxious desire to understand the prophetic parts of Scripture. It resulted in not a few being thoroughly and abidingly convinced, that whilst the Scriptures teem with predictions of blessing both to Israel and to the nations, previous to the dissolution of the present heavens and earth, yet that the period in which these promised blessings are to be dispensed, is one materially different from that in which the Church of God is now suffering; and that the great event which is to close this present dispensation and introduce the promised age of blessing, is the appearing in glory of the great God our Saviour, Jesus Christ.

I feel assured that I shall carry with me the feelings of every heart that has learned to expect

* This paper was first published in 1848.

the personal return of the Lord Jesus, when I say that they esteem it now as one of the most precious of the truths of God—at once the object of their hope, and the subject of their testimony; and I shall not in vain ask them to discountenance the circulation of any writings that tend to destroy the expectation of the Lord's return, or to throw any veil over the interpretation of those parts of Scripture that are connected therewith.

It grieves me to be obliged to number among such writings, the work of Mr. Fleming lately republished, and now extensively circulated. We almost instinctively feel a reverence for those who (though they may have been mistaken) have written and acted in the fear of God. This, however, must not deter us from openly exposing their errors; being persuaded of this, that they themselves, if they could now speak, would not ask that any thing they have said or done contrary to Christ and His truth should be spared. Their earnest desire would be, that every thing which through them may be acting prejudicially on the Church of God, might be frustrated in its results. And that must be prejudicial to the Church of God, which either subverts any of the great substantive facts that He has revealed, or which presents those facts in different order and connexion from that in which He has been pleased to set them in His word.

To those then who expect the pre-millennial

Advent of the Lord Jesus, (and to such principally I now write,) it is sufficient to say, that the book we are considering denies that Advent to be personal. On this ground only, I might ask that its claim should be rejected, for it claims to be regarded as a true exposition of the prophetic testimonies of God. *As such*, it must utterly be rejected. All who have learned to wait for God's Son from heaven, will feel too distinctly the solemn force of the concluding part of the nineteenth of Revelation, to admit of its being explained away on the ground of symbols or figurative language, or on any other ground. To such the quotation of the passage which I cite at the foot of this page would be sufficient to secure their condemnation of the work.* There are, however, a few other points to which I would desire briefly to direct their attention.

* "Seeing I have but slightly touched upon the millennium or the thousand years' reign of the saints on earth, I shall desire you to think a little further on this, as the greatest event that is to happen before the end of the world. I dare not indeed expatiate upon this vast subject; only I shall suggest a few things concerning it. The first is, that this is to begin immediately after the total and final destruction of Rome papal, in or about the year 2000; and that therefore Christ Himself will have the honour of destroying that formidable enemy, by a new and remarkable appearance of Himself, as I said before. But, secondly, we must not imagine *that this appearance of Christ* will *be a personal one*," etc.—*Fleming*, pp. 39, 93.

One of the great objects of God in giving enlarged acquaintance with the Scripture, especially prophetic Scripture, is to bring into His Church a more distinct recognition of the principles, which throughout this dispensation should have characterised His people, and marked them as separate, both from Israel, and from the nations. The leading nations of the earth, up to the very end of the present dispensation, are symbolised in the Scripture by fierce beasts which know not Christ and own no subjection to His laws; Israel is left to grope in judicial blindness and sin; but in contrast with both, the people of God stand peculiar in their principles and testimonies, commanded indeed to fight, but to fight only with spiritual weapons—holding both the shield and the sword, yet entering no path unsuited to those whose "feet are shod with the preparation of the gospel of peace."

Such, while the Apostles lived, were the principles and the position of the children of God. Such *ought* they to have been throughout the dispensation. But they soon abandoned this position, and mingling with the nations, learned their ways; and now as the latter days are closing in, God, whose word we have neglected, seeks, in mercy, to awaken us from our slumber, and to lead us back to that "sure word of prophecy" which marks the course of these nations, and their evil end. Many have read in the light of that word, the history and the doom both of Israel and the nations, and have seen the necessity of seeking to recover those principles, which give a characteristic separateness to the children of God.

Any book, therefore, especially any book on Prophecy, which counterworks this end; any book which tends to rivet the principles which have so long and so ruinously sunk the Church into identification with the nations, and deprived it of its peculiar testimonies, is most earnestly to be deprecated at a moment like the present. The principles of this book did not, and could not raise Mr. Fleming into dissociation from the nations or their ways. On the contrary, his principles (and it was their necessary result) led him to ask that their swords should be unsheathed for the maintenance or protection of the truths of God. A war between France and England is said to have been the immediate result of the influence he exerted over the monarch of the day. This was the practical result of his prophetic principles. How indeed could it be otherwise? For if angels coming forth with golden vials from the temple of God, and "clothed in linen pure and bright," are symbols really employed by God to designate the struggles and victories of Protestant over Popish nations, who would not desire to set such agency in action—who would not long to be engaged in conflicts worthy of being represented by symbols such as these?

But it is not so. They who are commanded to

follow Him, who, "when He was reviled, reviled not again," are not commissioned to wield the sword of destruction. If any thing could have justified its use by a disciple, it would have been the hour when Jesus was betrayed. But if even the defence of His sacred Person did not justify it—if even then it was said, "Put up thy sword within its sheath, for all they that take the sword shall perish by the sword;" how much more does it become us to allow it to continue sheathed till it shall please Another, holier and more mighty than we, Himself to take hold on judgment, and to terminate the age in which it has pleased Him to appoint that Truth should suffer. The very distinctive feature of the Church at present is meek and patient suffering. How then ought we to shrink from the principles of a system which destroys our apprehension of the Church's present calling; recognises not the distinction as drawn in the Scripture, between the coming dispensation in which Truth is to triumph, and the present in which it is appointed to suffer; and conceives that the sword of the earthly potentate is the fitting instrument whereby Christ's people are to be made to conquer.

But even if we could persuade ourselves to glory in those evil and cruel wars by which, after the true power and vigour of the Reformation had departed, the Protestant kingdoms fought for mastery; if we could deceive ourselves into the belief that the agency of earthly kingdoms contending with the Papacy, may fitly be represented by holy angels clothed in white, and that their triumphs were celebrated with thanksgivings in heaven; even if there were nothing in all this to shock us—if angels so clothed in white, could fitly represent the agencies of England, and Holland, and Sweden, contending with the Pope, shall we likewise say, that these same holy symbols are employed also to designate that agency which, in 1793, overthrew the throne of the French monarchs, and is now, in 1848, destroying, as is supposed, the authority of the Pope? This we must believe, if we receive the interpretations of Mr. Fleming. The triumphs of revolutionists and infidels, must be the triumphs of Christ and His truth.

I have no desire to shield Popery or its abominations. A system which solemnly and deliberately pronounces those who teach the blessed doctrine of justification by faith and the imputed righteousness of Christ, to be accursed, must be itself accursed.* (Gal. i.) But shall we, because we

* Conc. Trid. Canon IX. Si quis dixerit solâ fide impium justificari, ita ut intelligat nihil aliud requiri, quod ad justificationis gratiam consequendam co-operetur, et nullâ ex parte necesse esse eum suæ voluntatis motu præparari atque disponi, anathema sit.

"If any one shall say that an ungodly person is justified by faith only, so as to understand that nothing else is required to co-operate to the attainment of the grace of justification, and that it is in no respect necessary that he should see the wickedness of Popery, hallow, as it were, and pronounce blessed, the no less abominations of those agents of Satan who were permitted of God to overthrow the monarchy of France? Or shall we consider the triumph of those who are now

---

be disposed and prepared by the motion of his own will, let him be accursed."

Canon XI. Si quis dixerit homines justificari vel solâ imputatione justitiæ Christi, vel solâ peccatorum remissione, exclusâ gratiâ et caritate, quæ in cordibus eorum per Spiritum Sanctum diffundatur atque illis inhæreat, aut etiam gratiam quâ justificamur esse tantum favorem Dei, anathema sit.

"If any one shall say that men are justified, either by the mere imputation of the righteousness of Christ or by the mere remission of sins, to the exclusion of grace and love, to be diffused in their hearts by the Holy Spirit and to inhere in them, or moreover that the grace by which we are justified is only the favour of God, let him be accursed."

Canon XII. Si quis dixerit fidem justificartem nihil aliud esse quam fiduciam divinæ misericordiæ, peccata remittentis propter Christum, vel eam fiduciam solam esse, quâ justificamur, anathema sit.

"If any one shall say that justifying faith is nothing else than reliance on the divine mercy remitting sin for Christ's sake, or that that by which we are justified is reliance, and reliance only, let him be accursed."

Condemnation for the rejection of these the cardinal doctrines of our faith rests equally on that system known as Tractarianism in this country. I mention this, lest, because I maintain, as they do, that Antichrist is a secular person yet to arise, it should be supposed that I do not recognise the evil of their soul-destroying system.

striving to introduce the principles of infidel liberalism into Italy, to be indeed the victory of God's truth over the lie of Satan?* If it be so—if the agencies of such men as Voltaire and Robespierre—if liberalism and infidelity are to be thus regarded—if these agencies are to be symbolised by angels of white, and rejoiced over in heaven, then indeed we may abandon all hope of distinguishing between light and darkness, evil and good, the actings of God, and the actings of Satan. We may say that evil and good have exchanged places one with the other, or that all has been commingled in hopeless undistinguishable confusion.

The analogy of the past would little teach us, that that which overthrows evil, must itself be good. Jerusalem was overthrown for its iniquity, but were the nations that trampled it down blessed? Were they not, on the contrary, more wicked, more obdurate in evil, even than those whom they overthrew? Scripture and facts alike teach us, that it has been the order of God to allow evil to punish evil, and then to permit the evil that punishes, to exalt itself and to wax stronger in iniquity, than that which it has supplanted. And thus there has ever been an onward progress of evil, a strengthening and consolidation of iniquity; and prophetic Scripture declares that it shall be thus even to the end. But God has appointed that end; and this it is the especial subject of the Revelation to declare.

* This was written in 1848.

The object of the Revelation is not to trace the steps by which the evil of this present dispensation gradually advances. Its object is not the history of evil in its progress, but in its doom. It describes not *the growth* of the great and flourishing tree of Gentile greatness; but it declares the manner in which, *after it has attained that growth*, it will be smitten and cut down, and that by no mere human hand. The Revelation does not detail the steps by which one system of evil supersedes another, and then flourishes more abundantly than that which it has supplanted: but it reveals the form which the evil of this dispensation will exhibit when it has attained the maturity of its growth, and the manner in which God will at first chasten, and at last send His Son utterly to destroy that which has thus been allowed to ripen. This, however, has not been recognised by Mr. Fleming. He has read the Revelation as if it were the history of the *gradual* growth of evil, and of its *gradual* subjugation, by providential means. And even that evil he has sought in one small section of the prophetic earth, as if there were no evil against which God prophesies excepting Popery.*

* By the prophetic earth, I mean the part included within the Roman Empire. The Roman Empire embraced most important countries in the East; Syria for example, and Greece and Egypt. Its eastern division was more important even than the western. Yet by some strange carelessness,

And what has been the actual practical result of this system of interpretation? Has it placed the testimonies of the Revelation in firm and solemn opposition to the wickedness of these latter days? Has it put the sword of the Spirit into the hands of the Church for warfare against Satan, and enabled them to testify "against (επι) peoples, and nations, and tongues, and many kings?" No! It has done exactly the reverse. It has enabled the world, and some of the worst in the world, to seize on these most solemn of the testimonies of God, and to use them not merely as a palliation or defence, but even as a sanction, for the darkest deeds of their iniquity. At the time of the Revolution of 1793, Mr. Fleming's book was republished, both in America and in England, and used by the

---

the western only has been regarded by modern interpreters of prophecy, and the east forgotten. The seventh chapter of Daniel speaks of the ten horns (which represent the last ten kingdoms, into which the Roman territories are finally to be divided) as the horns of a symbolic Beast, which (as is admitted well nigh by every one) represents, not a part of the Roman Empire, but *all* of it. Nor does the second chapter of Daniel place the ten toes which represent the same ten kingdoms, on *one* only of the feet of the image. Five were found on each foot. We shall finally see the present convulsions in Europe terminate in the development of ten kingdoms, five in the eastern and five in the western branch of the Roman Empire. It is manifest that Popery has never prevailed in the eastern branch of the Roman Empire—but Antichristianism will.

liberal party of the day in justification of their views touching the occurrences in France. Doubtless it well suited them to be told that the revolutionists of Paris were symbolised by angels of God. But when we remember what the liberalism of that moment was, how truly it answered to the fearful picture which the Apostles draw, we may well tremble to think of the delusion which has made even the Book of Revelation to testify on their behalf. The Apostles speak of those who "despise government; presumptuous are they, self-willed; they are not afraid to speak evil of dignities. Whereas angels, which are greater in power and might, bring not railing accusations against them before the Lord. But these, as natural brute beasts, made to be taken and destroyed, speak evil of the things that they understand not; and shall utterly perish in their own corruption." (II Pet. ii. 10.) "Likewise also these filthy dreamers defile the flesh, despise dominion, and speak evil of dignities.* Yet Michael the archangel, when contending with the devil he disputed about the body of Moses, durst not bring against him a railing accusation, but said, 'The Lord rebuke thee.' But these speak evil of those things which they know not; but

* It would be well if many among the political Nonconformists were to lay to heart these words. They seem not to be conscious that Rulers are to be honoured as well as feared.

what they know naturally, as brute beasts, in those things they corrupt themselves." (Jude 8—10.) Such is God's description of the insubjection of the latter days. Such the iniquity which the book before us has been quoted to defend.

And now let us turn from the moral effects produced by this work to the system of interpretation adopted in it. It teaches (and in this, indeed, it is not singular) that "days" in prophetic Scripture do not mean "*days*," but "*years*," on which assertion the whole of this system is based.

As this assertion is so important, and affects so vitally every thought that we can form in connexion with the whole scope of prophetic enquiry, it becomes very needful to examine carefully the ground on which it is based. It would seem at first sight very strange, very contrary to the simplicity of Scripture, that God should use the word "*day*," when really He means not "day," but "year." What, then, are the grounds on which it is asserted that He does so?

The arguments used by Mr. Fleming are as follow: he refers first to Exodus xxiii. 10—12:— "Six years thou shalt sow thy land, and shalt gather in the fruits thereof. But the seventh year thou shalt let it rest and lie still, that the poor of thy people may eat; and what they leave, the beasts of the field shall eat. * * * * Six days thou shalt do thy work, and on the seventh day thou shalt rest; that thine ox and thine ass may rest, and the son of thy handmaid and the stranger may be refreshed."

These are the verses which Mr. Fleming quotes in support of his theory that a day means a year. But surely if there be any passage in which day means day, and year means year, it is this: for every one knows that Israel were commanded to keep every seventh year as a sabbatical year, and every seventh day as a sabbatical day; and every seventh day they *did* keep, and no one among them ever dreamed of its meaning anything else than a day. Throughout the passage "day" means "day," and "year" means "year."

The second argument is, that after the spies of Israel had searched the land for forty days, they were punished by wandering in the wilderness for forty years. Is day here put for year? Did they not search the land for forty literal days, and were they not punished for forty literal years? No passage could prove more convincingly that "day" means "day," and that "year" means "year."

Again, he quotes a similar passage in Ezekiel (Ch. iv.). Ezekiel is commanded to lie on his side for forty days, typically to bear the punishment of sins which Judah had committed for forty years. Did not Ezekiel lie on his side for forty literal days, and did not Judah sin for forty literal years? Ezekiel would surely have considered it a strange thing if he had been told that God had said indeed that he was to lie on his side forty *days*, but that He really meant forty *years!* It must have been so, if day be put for year. Did Ezekiel so think or so act? Did he lie on his side for forty *years?*

Mr. Fleming adds, "Nay, our Saviour Himself speaks in this dialect when He calls the years of His ministry 'days,' saying, 'I do cures to-day and to-morrow, and the third day I shall be perfected.'" But Mr. Fleming in saying this seems to have forgotten two things—first, that these words were spoken at the time when He wept over Jerusalem, and left it for the last time just three literal days before His death, whereas, on Mr. Fleming's theory, they ought to have been spoken during the first year of His ministry; secondly, he has forgotten that the duration of our Lord's ministry was not three years, but three years and a half. Each, therefore, of these arguments proves exactly the reverse of that which it is adduced to ustain.*

* Mr. Fleming adds also, "The seven years of Nebuchadnezzar's lycanthropy is called indefinitely *days* or *times.*" But this is not true. The seven years of Nebuchadnezzar's madness is never called "days." The Hebrew word for *days* is יָמִים and the Chaldee word for *times* is עִדָּנִין It is the same word that is used in the seventh chapter, where the power of Antichrist is said to be for a time, times, and dividing of time, *i.e.*, for three years and half a year, or 42 months, or 1260 days. Now, if three times and a half mean, as Mr. Fleming asserts, 1260 years, then Nebuchadnezzar's madness must last double that time, for it is said to be for

The next and last argument derived from Daniel's prophecy of the seventy weeks appears to an English ear more plausible. But it also utterly fails in proving the point intended. It appears more plausible to an English ear, because we are not accustomed to apply the word *week* (a word which our translators have chosen as their rendering of the Hebrew expression) to any period longer than seven days. We are not accustomed to say a *week* of weeks, or a *week* of months, or a *week* of years. We confine the expression to a seven of days. In Hebrew, however, it is otherwise. The word which our translators have rendered by "week" means simply what the Greeks would call a "hebdomad," *i.e.*, a septenary number, or a number consisting of seven. The word "*hebdomad*" stands in the same relation to "*seven*" that "*decad*" does to *ten*, or our English word "*dozen*" to *twelve*. As, therefore, we can say a dozen of days, or a dozen of months, or a dozen of years, so we can say a hebdomad of days (*i.e.*, seven days), or a hebdomad of months (*i.e.*, seven months), or a hebdomad of years (*i.e.*, seven years).

Now supposing that in the prophecy before us it

---

seven times, which would be equal, on Mr. Fleming's principle, to 2520 years, and consequently since he—*i.e.*, Nebuchadnezzar—lived only about 600 years before the Christian era, his madness must be continuing still. He must still be on earth: he must still be eating grass as oxen.

had been said, "Seventy dozen of *days* are to be fulfilled," and it was afterwards found that the prophecy really meant seventy dozens of *years*, then indeed it might be said that "*day*" meant *year*, and the point would be unquestionably proved. But if the word day was not mentioned at all, and the prophecy simply said, "Seventy dozens are to be fulfilled," then we should say that the prophecy was ambiguous—that it might mean dozens of days, or of months, or of years, and that we must endeavour by other means to discover which of these was intended. And if on examination we found that *days* should be supplied, we should supply days, and interpret them as days; or if *months*, we should insert months, and interpret them as months; or if *years*, we should insert years, and interpret them as years. In either case, day would mean day, and month, month, and year, year. It would simply be a question of which should be inserted. There would be no question respecting their meaning when inserted.

Thus is it in this prophecy of Daniel. It is not said, "Seventy hebdomads *of days* are appointed;" it is merely said, "Seventy hebdomads are appointed." Consequently, seeing that the word "day" does not exist in the passage, that which does not exist cannot be put for any thing, nor mean any thing, and there is an end of the question.*

---

* See this passage respecting the "seventy hebdomads" considered in "Prospects of the Ten Kingdoms," page 215.

Such, then, are the reasons given for this strange imagination—reasons of which it may be fairly said, that they are no reasons at all, and therefore it follows that the whole superstructure built on such a foundation utterly falls.*

* Some have said, that the Jewish Rabbis teach that "day" is to be understood as meaning "year." It is true indeed that Rabbinism has enough to answer for as respects perversion of Scripture, but it is not amenable to the present charge. The Jews do not allow that day is put for year. Aben Ezra, who lived in the twelfth century, says, "Our honourable master Saadias expounds correctly and well, in Holy Scripture days are always days, and never years. Yet it is possible that the word 'day' may mean an entire year, since the repetition of the days produces a return of the year, as when it is said in Ex. xiii. 10, 'from days to days,' *i.e.* 'from year to year'; days meaning a complete year. *But when the number is stated as two days, three days, it cannot mean years, but must be days as it stands.*" (See Aben Ezra as quoted by Maitland in his Apostles' School of Prophetic Interpretation.)

"The calculation," says Dr. Todd in his sermons preached before the University of Dublin, "from which Mede has derived his main position, that 'the time of the end,' or the coming of the Antichrist, began in the twelfth century, depends altogether on the untenable assumption that *days* in prophetic language denote *years;* an assumption, which an eminent living writer has so completely refuted, that no theory built upon it can be considered as requiring any further confutation. I shall not therefore detain you by repeating the arguments employed by the writer to whom I allude and other learned men, to overturn this principle; it may be enough to remark, that every prophecy, which is already known to have been fulfilled, and in which days or years are

And if it be asked, how it can have happened that Mr. Fleming's predictions have in certain cases been verified by the event, I reply, where is there any instance of such verification? The verification of a Scripture prophecy is the fulfilment of it *in that sense in which God has spoken it.* Where then has any prophecy of the Scripture been fulfilled at the time, and by the circumstances spoken of by Mr. Fleming as about to fulfil it? Have circumstances which he has predicted fulfilled one prophecy of the word of God? We may safely answer, No!

Let us take for example his prediction respecting the present year, 1848,* in which he supposes the 1260 days will end by the weakening of the Papacy under the pouring out of the fifth vial. Now supposing that 1260 days mean 1260 years, which they do not, and supposing the Beast symbolised the Papacy, which it does not, and supposing the previous vials had been poured out, which they

---

mentioned, can be shown to have been fulfilled by literal days, and literal years; and that the opinion that a day, in prophetic language, means a year; and a year, three hundred and sixty years, is an arbitrary assumption, destitute of any Scriptural evidence, and which, even those commentators who most warmly advocate it, are unable uniformly to adhere to." (Discourses preached before the University of Dublin, at the Donellan Lecture in 1838, by James H. Todd, B.D., Fellow of Trinity College, Dublin, p. 19.

* The first edition of this book was published in 1848.

have not been—yet even then, will any one venture to affirm that Rome has in this year begun to be full of darkness, in the sense meant in the Scripture, and that they have begun to "gnaw their tongues for pain, and blaspheme the God of Heaven, because of their pains and their sores?" Revelry and rejoicing, because of conquest and deliverance, is far more likely to resound from one end of Italy to the other. Where then is there any fulfilment in this? But again, his prediction is, that the 1260 days are to conclude in 1848 by the Papacy being *weakened*, not *abolished*. He denies that it can be *abolished*. Now it is impossible that this prediction can be fulfilled, either in this year, or any other year. It is a prediction that secures its own frustration. For whenever the period of 1260 concludes, (and, as regards the present point, it is immaterial whether it be a period of days, or years,) whenever it ends, the reign of evil ends from one end of the earth to the other. The day of man ends and the Day of God begins. The dominance, not merely of Popery, but of every other evil system in the earth, will then utterly, and for ever terminate. The night will have passed, and the morning without clouds will have arisen : it will have dawned with its unchangeable light of blessing on a reconciled and recovered earth. His prediction, therefore, seeing that it consists of two parts necessarily incompatible with each other, secures its own frustration. If the Papacy is only to be weakened, and not abolished, then his prediction respecting the termination of the 1260 days is falsified; for they cannot terminate until all evil terminates. If, on the other hand, the period of 1260 were now to end, then his prediction respecting the Papacy being weakened merely, and not abolished, would be falsified; for it would be abolished, and not weakened.

Whether the Papacy be really weakened by that which is now occurring at Rome, is a question on which I express no opinion. Facts may possibly prove that it is rather being strengthened. But however this may be, we may with all certainty affirm, that the events now happening at Rome are no more the fulfilment of the passage above quoted in the Revelation than the conversion of Constantine was the fulfilment of the solemn words that follow the opening of the sixth seal. He who can persuade himself to believe, that a vision such as that in which " *the sun became black as sackcloth of hair and the whole of the moon* (ἡ σελήνη ὅλη) *became as blood; and the stars of heaven fell unto the earth, even as a fig-tree casteth her untimely figs, when shaken by a mighty wind. And the heaven was separated from its place as a scroll when it rolleth itself together* (ἀπεχωρίσθη ὡς βιβλίον ἑλισσόμενον) : *and every mountain and island were moved out of their places. And the kings of the earth, and the great men, and the chief captains, and the rich men, and the mighty men, and every bondman, and freeman, hid themselves in the caves and in the rocks of the mountains; and they say to the mountains and the rocks, 'Fall on us, and hide us from the face of Him that sitteth on the throne, and from the wrath of the Lamb: because the great day of His wrath hath come; and who is able to stand?'"*—he who can persuade himself, that a vision like this has been accomplished in such an event as the nominal conversion of the Roman Empire, will certainly find little difficulty in making any part of the Word of God accommodate itself to any prediction that he may please to utter. Such interpretations of the Word of God will accommodate themselves to anything, and mean anything. They are simply *non-natural* interpretations, false meanings put upon the *words of God*. Mr. Fleming has accepted this interpretation of the vision given at the opening of the sixth seal. It is not wonderful, therefore, that such a system of interpretation adopted throughout the Revelation as a whole, should open a door sufficiently wide for all kinds of imaginary fulfilments. But there is in all this no accomplishment of the Word of God.

To take another example, viz., his conjecture respecting the fulfilment of the fourth vial in the abasement of popish countries throughout the seventeenth and eighteenth centuries, and especially by the overthrow of the French monarchy in 1793. Whatever may be thought of this as a conjecture, it has no title to be regarded as any fulfilment of

Scripture prophecy. The words of the Scripture are these, "*And the fourth angel poured out his vial upon the sun; and it was given unto it to scorch men with fire. And men were scorched with great scorching, and blasphemed the name of God, who hath authority over these plagues: and they repented not to give Him glory.*"

Now even if we were to admit that the sun represents a number of popish monarchs and kingdoms, from our James II. to Louis XVI. of France, how is it that increased power given to the sun indicates increased weakness in those whom that sun is said to represent? If these kings and kingdoms are symbolised by the sun in this passage, we ought to have seen them, and all popish kings and kingdoms, greatly increased in power, and beginning to scorch others, (I suppose) Protestants; for they could hardly be supposed to scorch themselves. This surely must have been the interpretation—the Papists must have been strengthened, not weakened; yet for some unexplained reason, Mr. Fleming suddenly reverses the symbol and says that *the strengthening* of the sun indicates *the weakening* of the Papists, whom that sun is said to represent. Is this fulfilment of prophecy?

But again, when James II. was driven from England, or when Louis XVI. fell, or during the intervening centuries, were men so "scorched with fire" and "great scorching," or did they in an especial manner, because of their special plagues, blaspheme God? The fall of James II. is looked on in England as the dawn of liberty and prosperity; and the fall of the French monarchy is looked at in France as issuing in the most glorious period of their history, when under Napoleon they shivered Europe under the Empire. As to the Papists, their joy and self-congratulation at the supposed advance of their system, both in France, England, and the Colonies, is certainly very unlike the wail of torment or despair.

It would be easy to accumulate instances of similar inconsistencies of interpretation. We might ask, If the ten-horned beast of Revelation xiii. represents the Papacy, how does it happen that there are within the sphere of its influence unconverted men who are *not* Papists? For (with the exception of those whose names are written in the Lamb's book of life) all who fall within the scope of "the Beast's" influence will worship him when he appears. Have all wicked men worshipped the Pope? We might ask how, if the ten-horned beast represent the Pope (who is ecclesiastical), *another* beast, also ecclesiastical in character, is in the same chapter (Rev. xiii.) said to minister in his presence, 'and what the image is, which speaks and is worshipped,' and commands that whosoever will not worship it should be killed? Has the world ever yet seen such an image? If so, where is it? And where are its worshippers? We might ask, too, how a beast, which wears throughout the whole of the 1260 days the crowns of *all* the kingdoms of the Roman world, can represent the Pope, who has never worn those crowns? And how, if the harlot represents the Papacy, can she be destroyed by the ten horns of that beast (Rev. xvii. 16, 17), which is supposed to represent the Pope? How is it that during the whole of the 1260 days the beast is represented as having authority in the holy city, even "the city where our Lord was crucified" (see Rev. xi. 7, 8), whereas the Pope has never exercised authority in Jerusalem at all? We read that all who shall worship the beast or receive his mark are certainly doomed to everlasting wrath (Rev. xiv. 9), whereas it is possible for Papists to believe and to be saved. Many a question of this kind might be multiplied, difficult indeed to be answered by those who seem to speak as if all evil were concentrated in the Pope alone.

God does not think so. He sees in the corrupt churches of the East, in Mahomedanism, in Judaism, and in *latitudinarian* Protestantism, evil as great, and, in the latter instance, greater than in Popery itself; and He has not delineated in His word the final climax of evil, without including in that delineation all the channels which unite in that final meeting point. The Old Testament, in many of its prophecies, describes the same period as that of which the Revelation treats, that is to say, the period which immediately precedes the appearing of the Lord. It speaks abundantly of the antichristianism of Israel in Jerusalem, and of the antichristianism of the great Gentile nations (the nations of the Roman earth) at the closing hour. Does the Revelation contradict these testimonies of the Old Testament Scriptures, or is it silent respecting these great events of the latter day, which are to knit the East and the West together in a confederacy of evil, the like to which has never yet been seen? Is it silent respecting the very events which give to the period of which the Revelation professes to treat, its most distinguishing and characteristic features? It must be silent about these things if it only speaks of Popery; for neither the Jews, nor half of the Gentile nations are Papists, and therefore, never could be included in the warnings respecting the beast or his image, if Popery be that which is designated thereby.

And is this really to be the burden of our testimony? Are we to teach the Jew, the Mahomedan, the Protestant infidel, that he is in no danger from that "hour of temptation that shall come upon the whole world, to try those who dwell upon the earth." That hour will bring with it a mightier strength of delusion, and a power of judicial blinding more terrible than any that has ever yet visited the earth. Are Papists the only persons that will be endangered by it? (Rev. iii. 10.) Are we to teach them that Popery so concentrates within itself all evil, that they who are not Papists are safe from the fearful denunciations of the closing book of the Scripture? Are we thus to nullify the word of God—to turn the course of the sword of the Spirit, and to blunt its edge?

This no doubt is the desire of Satan. To lead into such vagueness of thought and interpretation as takes all definiteness from the symbols and expressions of Scripture; to confine the most solemn of its warnings and denunciations within a circle too narrow to include half of those for whom they are intended; to persuade us that that is past, which is not only future, but near; and so to hide from men the gulf which is yawning almost before their feet—this no doubt, is one of the chief anxieties of the great enemy of souls. And he has fearfully succeeded. He has succeeded in perverting, when he has not been able to hide, the very light which is set to guide us through the night until the day-star arise.

No one, I suppose, who believes in the personal existence of Satan, will doubt that he is able to tempt, and to put evil thoughts into the heart. And if able on other subjects to deceive, he is able also to lead into wrong interpretations of the word of God. He is acquainted with the facts that Scripture reveals as about to be, and he is frequently himself permitted (as when he was allowed to afflict Job) to conduct the train of circumstances which is to lead on the series of events to their appointed issue. It is not therefore to be wondered at, that, after having turned our minds into a wrong channel of thought, he should suggest such anticipations as should receive confirmation from the events which he knows are about to happen.

If for example, we could now be persuaded to believe that the waning of the Turkish power is the fulfilment of that passage in the Revelation which describes the pouring out of the sixth vial on "the great river, the river Euphrates"; if we could be seduced into the belief that when the Scripture speaks of "the way of the kings *from* (ἀπο) east" being prepared, it refers to the return of Israel to their own Land; if the words, "Behold I come as a thief; blessed is he that watcheth," etc., are not to be understood of the personal advent of our Lord, but indicate merely some providential interference of His hand—if we could be induced to accept interpretations such as these, we should not be long in finding in events soon about to occur, an apparent verification of our expositions. It is true that the sphere of Constantinopolitan power and influence will be narrowed greatly. Syria for example will be separated from Constantinople, just as Egypt and Greece have been. That might be represented to be the pouring out of the sixth vial on the Euphrates. Israel, while yet unconverted will return to their own Land. This will be regarded as the coming of the Kings from the East, an apparent verification of our expositions; and the false prosperity which for a season will attend them after their return, will be represented as the figurative coming of the Lord. We might even venture to predict the periods (for they are not far off) when these things are likely to be, and yet all would be delusion. The events are not *the* events which God has intended to indicate, and the very fulfilment of our anticipations, would only lead into deeper darkness as to the real meaning of His word.

There are many other grave and serious objections that might be urged against this work. It might be shown how entirely it fails in *rightly dividing* the Scriptures. It is really written as if God had made no distinctions in His word between the Jews, the Gentile nations, and the Church. The history of Christianity is confounded with that of the Gentile *nations:* and as to Israel, their history, both when they return to their land in unbelief, and also after they are restored under the blessing of the Lord—all that they are to be in sin, and all that they are to be in millennial blessedness, appears to have been unnoticed and unknown. And yet Israel is the very centre of the earthly arrangements of God; the turning point on which His dispensations hinge.

But I have said enough. He who, after considering the practical effects of this work—its failure in the very first principles of prophetic interpretation, and above all, its rejection of the personal Advent of the LORD, can yet consent to regard it with complacency, will be little affected by any accumulation of further evidence. But I trust that many will escape the snare; and that the very publication of this book may tend to excite an enquiry that may end in the discovery of truth. The very perplexities in which this subject has been so industriously involved, may teach us its importance, for in a world like this, that which is most precious, is likely to be most hidden. But if there be a little patience—a little honest perseverance in using the Scripture, a real desire "to buy Truth and sell it not," these mists of darkness will disappear, and truth will be seen in the clearness and simplicity which might be supposed to belong to a subject, which the word of GOD professes to REVEAL. "His testimonies are sure, making wise the simple."

CHAPTER XI.

THE PROPHETIC SYSTEM OF MR. ELLIOTT AND DR. CUMMING CONSIDERED.

THE political and social convulsions which spread throughout Western Europe after the late French Revolution in 1848, had a marked effect in arousing the minds of many to enquiry respecting the future. They knew that they were very ignorant of all that God had revealed touching the prospects of the nations: they felt ashamed of that ignorance, and not a few rushed with inconsiderate haste to the perusal of any writings that seemed likely to satisfy their enquiries. The avidity with which the book of Mr. Fleming was sought after, bears witness to the state of feverish excitement that then prevailed. Two years have since passed, and now that book is scarcely heard of any longer. Men probably are convinced that Mr. Fleming was wrong. They see that in 1848, Popery neither received its death-blow, nor ceased to be supported by France. Indeed, the attempt now being made in Austria, Italy, and France, is to rule by help of the Jesuits; and with this object, Popery is sustained. The final impression left on men's minds by Mr. Fleming's work, will probably be a painful conviction of having been deceived; and thus deeper prejudice than ever will be excited against every form of Prophetic enquiry.*

There is, however, another work, the *Horæ Apocalypticæ* of Mr. Elliott, which although not less open to objection than that of Mr. Fleming, is still maintaining a considerable influence over many. Its influence has been extended more widely, in consequence of its having been made the groundwork of a popular exposition of the Revelation, lately published by Dr. Cumming. Dr. Cumming avowedly adopts the opinions of Mr. Elliott, stating

* Many changes have taken place since this was written, all tending to frustrate the predictions of Mr. Fleming. During the reign of Napoleon III. the Papacy was vigorously upheld by France; and even now its power in that country, though curbed, is not destroyed. In England the influence of Romanism has greatly increased. Nevertheless, neither in France, nor in England, nor in Austria, will Rome again become *dominant*. In all the Latin half and in all the Greek or Eastern half of the Roman World none of the old religious systems will regain their ancient *supremacy*. The conflict between them and Latitudinarianism may be fierce, and perhaps prolonged, but it will end in the triumph of Latitudinarianism, which subsequently will have itself to succumb before the supremacy of the Antichrist. In those parts of Christendom which are external to the Roman boundary the doctrinal, if not the secular, system of Popery will probably have great influence until the end. Not till then will Iniquity hide its head.

that he regards his "Horæ" as about to "occupy a place in reference to unfulfilled prophecy, that Newton's 'Principia' has occupied in reference to science."* We may therefore consider these two writers as identified. In the following remarks, I shall regard them as one—and my observations, even when I do not specify the name of each, must be regarded as applying equally to the opinions of both.

It will not be necessary, in order to determine the character of these works, to follow the authors through all their statements. The consideration of a few cardinal points will be sufficient; nor will I at present enter into the question whether the Revelation be intended to supply the continuous history of the past 1800 years, or whether it be on the contrary, the declaration of the manner in which God will visit with His judgments the evil of the present dispensation *when matured.* Mr. Elliott and Dr. Cumming have unhappily determined, that the Revelation *does* supply the consecutive history of the past 1800 years; and hence their effort to find in it the record of events, which the very character of the Book requires to be passed over in silence.

One of the most remarkable events that have occurred during the past 1800 years—an event which Mr. Elliott supposes to be largely dwelt on in the Revelation, is the assumption of the pro-

* Cumming, p. 15.

fession of Christianity by the Roman Empire in the days of Constantine. Long before Constantine, not only had the freshness and vigour of Christianity entirely departed, but there was scarcely a truth that had not been adulterated or denied— scarcely an error that has since flourished under Popery, of which the seed had not been abundantly sown. Even the very framework and order of the Church had become entirely diverse from that which the Apostles had prescribed in their Epistles. Persecution instead of driving back to God, caused many, according to the words of the Lord, to be offended—they betrayed and hated one another—iniquity abounded, and love waxed cold. (Matt. xxiv.) Well indeed it may be said that iniquity did abound. The Church had become in many places, a mere school of philosophy—an arena for the display of Rhetoricians and Sophists: and the writings of those Teachers who were considered most orthodox and influential, such for example as Origen, had deepened the darkness which since the Apostles died, had steadily settled in upon the word of God. Heresies of various kinds abounded; and Arianism, at the time when Constantine assumed the profession of Christianity, was in many places destroying the very form of orthodoxy.

Such was the condition of the fallen Church, when Constantine, professedly converted by a vision, adopted the Cross as his distinctive military ensign. That token of the relation of the Prince of

Peace to a world that had rejected Him—the sign to a believer that he has been crucified to the world, and the world unto him—the sign also that he has been entrusted with the ministration of the Gospel of peace and grace—that emblem was now adopted by the proud and warlike master of the Roman World as his peculiar device. "It was seen," says Mr. Elliott, "glittering in the helmets, engraved on the shields, and interwoven with the banners of his soldiers. More especially in his principal banner, the *Labarum*, he displayed at its summit the same once accursed emblem, with a crown of gold beside it, and the monogram of the name of Him who, after bearing the one, now wore the other." (Elliott, p. 113.)

The character of the religion of Constantine may be judged of from this, that subsequently to the celebrated vision by which he was said to be converted (A.D. 312) he publicly invoked the Deity as one and the same in all forms of worship: and at a later period (A.D. 321), he promulgated simultaneous edicts for the observance of Sunday, and the due consultation of the heathen soothsayers.* Many of the heretics and schismatics, whose conventicles he suppressed, "through fear of the Imperial threat," says Eusebius, "joined the Church with hypocritical minds." The prelates of the Church began to revel in king's courts. Eusebius

---

* See Newman's Hist. of Arianism, p. 281.

Bishop of Nicomedia, and Eusebius Bishop of Cæsarea, were Constantine's chief favourites. Both were patrons of Arianism, the former probably because he virtually was an Arian, the second because he was a worldly latitudinarian. The first banishment of Athanasius, and the temporary restoration of Arius, was effected by their wiles; nor did even the awful death of Arius under the hand of God, damp the efforts of the court party against those who still retained the true confession of the Person of the Son of God. Constantius, the successor of Constantine, was a persecuting Arian.*

* Immediately after the Nicene Council, the semi-Arian party, who hypocritically adopted its confession, contrived, by flattering Constantine, and accommodating themselves to his political and latitudinarian views, to get the power into their own hands ; and the orthodox, after having apparently triumphed at the council, found themselves discountenanced and persecuted. This was one of the first results of the Church being patronised by the Imperial power. Sulpitius Severus, born A.D. 366, who was by no means disposed to look unfavourably on the attempt to dignify Christianity, after referring to what the Empress Helena had done in building churches, and the liberty from persecution attained under a Christian Monarch, proceeds thus : " But a far more tremendous danger ensued to all the Churches from the peace thus acquired, for at that time the heresy of Arius burst forth, and by the error it introduced threw the whole world into confusion. For even the Emperor himself was drawn aside by the two acute authors of this perfidious doctrine, both of whom bore the name of Arius. (Etenim duobus Ariis, acerrimis perfidæ hujus auctoribus, Imperator etiam

This is but a faint picture of the corruptions of that evil hour. How then can it be imagined, that *such* an exaltation of *such* Christianity could be

---

depravatur.*)* Under the impression that he was fulfilling a religious duty, the Emperor began violently to persecute; the Bishops were driven into exile; the clergy were raged against; the laity punished in cases where they had separated themselves from communion with the Arians. The doctrine which the Arians maintained was to this effect, that the Father of the Lord, for the sake of creating the world, had begotten a Son—that so according to His almighty power, a new and another Lord was made, in a new and another substance; and that a time was when the Son was not. In consequence of this evil, a Synod was convened at Nicæa from the whole world. Three hundred and twenty-two Bishops were assembled; a full confession of faith drawn up, and the Arian heresy condemned. The Emperor accepted the Episcopal decree. The Arians, not daring any longer to treat of the question in opposition to the orthodox faith, mingled with the Churches, as if acquiescing in the decision, and not differing in opinion. Nevertheless, there remained in their bosoms a deep-seated enmity against men of catholic doctrines, and seeing that they could not overcome them in discussion, they began to attack them by suborning accusers and inventing charges. * * * By such means they so far influenced the Emperor that Athanasius was sent into exile to Gaul." (Sulpitius Severus, Lib. II., c. xxxvi.)

As regards Eusebius, nothing can be more wickedly latitudinarian than the manner in which he continually speaks of the Arian controversy. Instead of marking the doctrine of Arius, as being, what it really was, utterly subversive of the faith, he strives, in his life of Constantine, to represent the controversy raised upon the question as

the work of God in blessing? Besides, even if there had *not* been such corruption of truth, could it ever have been the desire of God that the Cross should be identified with the banner of Imperial Rome— setting aside for the present the question of the ungodliness with which that banner was associated? If the real, true, pure doctrine of the Cross had verily become supreme among the nations, the essential character of the Dispensation must have been

---

entirely needless, and speaks of the Bishops on both sides as "arranging themselves in angry hostility against each other on pretence of a jealous regard for the doctrines of Divine Truth." (Ch. 61.) He quotes with delight the letter which Constantine wrote to Alexandria, in which the Emperor speaks of the difference between Arius and his opponents as being "intrinsically trifling, and of little moment" (μικρας καὶ λιαν ευτελους) and tells them that they are in reality "of one and the same judgment."

It is said by some that Constantine would not have so written after he had learned more about the real nature of the controversy. That is questionable; for he persecuted Athanasius after the Nicene Council. But, however this may be, it cannot affect Eusebius. Eusebius well understood the controversy, and wrote thus, long after the Nicene Council—and Eusebius was Constantine's favourite. The condition of Eusebius' mind on other subjects may be judged of from the following passage. Speaking of the structures that had been reared by Constantine at Jerusalem in honour of Christ, he adds, "It may be that this was that second and new Jerusalem spoken of in the predictions of the prophets, concerning which such abundant testimony is given in the divinely inspired record." (Eusebius, Vita Constant., ch. xxxiii.)

changed. The time would have come for the saints to reign; and Constantine the Master of the Roman World, and the Prelates who shared his dignity, would have become the proper representatives of the right condition of Christian Truth, instead of the Apostle Paul who died daily. But was it so?

Mr. Elliott and Dr. Cumming according to their principles, should answer, Yes.*

It is said that the Cross was seen by Constantine in a dream, with the inscription, "By this conquer." But is it forgotten that Satan can send deluding dreams, and that God is likely to permit it, when men cease to walk by the principles of His written word, and give themselves over to other guidance? Or, dreams may arise from the multitude of our own thoughts within us. In every case, both miracles and dreams need to be tested by the word of God. Their origin is soon determined to be not from Him, if they lead into practices which His word condemns. There had been another occasion on which Satan had sought to connect himself with Christianity and to grasp its influence. When Paul and Barnabas first turned their steps towards Europe, and began to preach in Philippi, who would have

---

* "Constantine (says Mr. Elliott) was the first *crusader;* and, with better reason than the Princes of the eleventh century at Clermont, might feel as he prosecuted the war that it was 'the will of God.'" (Elliott, p. 114.)

thought that an unclean spirit would have been the first to say of those two despised strangers, "These men are the servants of the Most High God, who show unto us the way of salvation?" Satan's object, doubtless, was to connect himself with the testimonies of Christianity, in order that he might mar, disgrace, and ruin them. St. Paul was on his guard; but the Church in the days of Constantine were not as St. Paul. They had long ceased to be guided by the word of God. They had long coveted exaltation.

It is not necessary to say whether the vision or dream of Constantine came immediately from Satan, in order to determine the character of the Christianity that prevailed both at, and after, his conversion. That is a question for *conscience* to decide. And will any real Christian with the Scripture in his hand, deliberately say, that even the orthodox Christianity that prevailed before Constantine, was such as God could honour: or, if He had intended to honour it, that He would have done so by exalting it into the Imperial Throne? This is a question which needs no reference to prophetic Scripture: it is determined by the essential characteristics of the primary doctrine of Christ.*

If the conversion of Constantine and the subsequent exaltation of Christianity, had really been

* Hagenbach, in his "History of Doctrines," gives the following account of doctrines taught immediately after the

an act of God's hand in blessing, there are two parts of Scripture, which, from the nature of their

---

Apostles died. It shows how early the distinctive truth of the Gospel was lost.

Clement, of Alexandria, writes as follows: "Perhaps as we have been redeemed by the precious blood of Christ, so some will be redeemed by the precious blood of the martyrs." (Vol. I., p. 178.)

"Origen also ascribed somewhat of the effects of an atonement to the death of the martyrs, as Clement had done before him." (Hagenbach, p. 173).

"That theory of satisfaction had not then been formed which represents Christ as satisfying the justice of God, by suffering, in the room of the sinner, the punishment due to him. The term 'satisfactio' occurs indeed in the writings of Tertullian, but in a sense essentially different from and opposed to the idea of a sacrifice made by a substitute." (P. 173.)

"The forgiveness of sins was made dependent both on true repentance, and the performance of good works." (P. 180.)

"From all that has been said in reference to the subject in question, it would follow that the primitive Church held the doctrine of VICARIOUS *sufferings*, but not of *vicarious satisfaction*" (p. 178)—that is, although they admitted that no one could be saved unless Christ had suffered on his behalf, yet they also taught, that something else was needful in addition to those sufferings of the Redeemer: consequently, His sufferings did not make "*satisfaction*." Thus the distinctive truth of the Gospel was utterly set aside, and the doctrines of the Epistles to the Romans and to the Hebrews virtually blotted out from the Bible.

What a mercy to be able to turn from these "doctrines of men" to the sure testimonies of God's own word. There

subject, must have so mentioned it. One is the prophecy of our Lord in the 24th of Matthew. In that prophecy the Lord speaks of the early persecutions which Christianity would meet when it should go forth among the Gentiles. "Then shall they deliver you up to be afflicted, and shall kill you: and ye shall be hated of all the Gentiles (ὑπο παντῶν τῶν ἐθνῶν) for My name's sake. And then shall many be offended, and shall betray one another, and shall hate one another." Such is the picture of the condition of Christianity in the midst of its early persecutions. And does the Lord say, that after this the scene would change, and that God would interfere to exalt Christianity in the earth and so give it His blessing? No: the prophecy says exactly the reverse; it says that "iniquity shall abound, and the love of the greater part (τῶν πολλῶν) wax cold," so that it should be a hard thing to "endure unto the end." The Lord could not have thus spoken, if God had miraculously interfered in the days of Constantine to make Christianity triumphant. Again, in the seventh of Daniel, the Empire over which Constantine ruled, is represented *from the beginning to the end of its course*, by one unaltered symbol, viz., a "beast dreadful and terrible, and

---

we find that all who believe, are "sanctified by the offering of Christ's body once" (Hebrews x. 10)—and that all so sanctified are, as to acceptance, "PERFECTED FOR EVER." (Heb. x. 14.)

strong exceedingly;" always evil, and, at the last, blasphemous. It would have been impossible to represent it thus, if God had at any one moment of its course, brought it into holy and blessed connexion with the Cross of His Son. In that case the symbol would have been changed. The terrible monster would have disappeared, and Constantine's Empire would have been represented by a Lamb, or some other such symbol of gentleness and peace. But the symbol remains unchanged. The Roman nations governmentally regarded (whether Pagan as at first; or nominally Christian as under Constantine, or as now, partly Mahommedan, partly of the Greek communion, partly Papal, and partly Protestant) are, in each and in all these conditions, represented by the same Monster which ends its course in blasphemy, and is then adjudged to be "given to the burning flame," and the "time came for the saints to possess the kingdom." (See Dan. vii. 22, and Rev. xi. 17.)

And now let us consider those parts of the Revelation which are interpreted by Mr. Elliott and Dr. Cumming of the conversion of Constantine. The first is the following most solemn passage from the end of the sixth chapter. "And I saw when he opened the sixth seal, and there was a great earthquake; and the sun became black as sackcloth of hair, and the whole of the moon became as blood; and the stars of heaven fell unto the earth, even as a fig-tree casteth her untimely figs, when shaken by a mighty wind. And the heaven was separated from its place as a scroll when it rolleth itself together; and every mountain and island were moved out of their places. And the kings of the earth, and the great men, and the chief captains, and the rich men, and the mighty men, and every bondman, and freeman, hid themselves in the caves and in the rocks of the mountains; and they say to the mountains and the rocks, 'Fall on us, and hide us from the face of Him that sitteth on the throne, and from the wrath of the Lamb: because the great day of His wrath hath come; and who is able to stand?'"

This is the passage which Mr. Elliott and Dr. Cumming interpret of the political fall of Paganism in the days of Constantine. Mr. Elliott, indeed, seems constrained to allow, that the language may also refer to the great final convulsion when the Lord shall return, and shake all things; and adds, that it is the habit of Scripture to make a less event *of the same character*, typical of a greater that is to follow. It is indeed most true, that a less event *of the same character* is often made to *foreshadow* (I would not say be *typical* of) a greater that is to follow. Thus the past destruction of Jerusalem, forewarns us of the greater and more terrible season of tribulation yet to come on that City. The desolations that have fallen upon Egypt and the cities of the East, are intended to warn us of greater judgments yet to fall upon those regions.

In such cases, the actings of the Divine hand may rightly be said to be *of the same character*. But are we to believe that the great final act of God in shaking all things in order to establish the glorious kingdom of His Son, is *of the same character* as that by which worldly secularised Christians, whom He designed to chasten, were permitted by Him to introduce the wicked Christianity of the days of Constantine? Did God indeed put forth His mighty power in order to establish and give effect to evil? Is He the author of sin?

It would be strange, indeed, if a period when all kinds of worldliness flourished with new strength, and when even Paganism continued in such vigour as after a little again to threaten Christianity with extermination—it were strange if such a period were one in which the great day of the Lamb's wrath had come, or in which men imagined that it had come. There were, doubtless, many individual Pagans who were vexed and angered that Christianity was favoured, and Paganism endangered; but does Mr. Elliott really believe that there was one living Pagan throughout the whole Roman Earth, who thought, when Constantine assumed the profession of Christianity, that the day of the Lamb's wrath was come? Why, they did not not believe even in the existence of the Lamb of God. How, then, can it be said that "EVERY bondman, and freeman," as well as "the kings of the earth, and the great men, and the chief captains, and the rich men, and the mighty men, hid themselves in the caves and in the rocks of the mountains; and said to the mountains and the rocks, 'Fall on us, and hide us from the face of Him that sitteth on the throne, and from the wrath of the Lamb?'" It is language thoroughly inapplicable to the days of Constantine, for then not only was there no general wail, but the shield of Imperial protection was thrown even over Paganism itself. It was anything rather than a day of general woe.*

The disastrous effect of explaining away passages like this, which belong exclusively to the great coming hour of Christ's return in glory, and the encouragement thereby afforded to neologian tendencies, is most manifest. Two things should have been sufficient to guard Mr. Elliott against such an interpretation of this passage, first, a conscientious judgment of the circumstances, which would have showed him that the establishment of

* The twentieth anniversary of Constantine's accession (says a recent writer) was hailed with general rejoicings throughout the world. Since the Augustan age, so long a cessation from bloodshed had not been known. "Throughout the whole kingdom," says Eusebius, "the sword of justice hung idle; and men and nations obeyed rather from filial affection than by means of force." The heathen rejoiced scarcely less than the Church: it was clear that no rod of iron was as yet to be swayed by the triumphant faith. Therefore, on the twentieth anniversary, the people set apart a day for feasting. (Maitland, p. 207.)

the wicked Christianity of the days of Constantine, could not have been an act of God analogous to that by which He will introduce the kingdom of His Son. Secondly, a regard to the structure of the Book of Revelation would have preserved him from the error. It is written on the same principle as the visions of Daniel — visions which are not chronologically successive to each other, for the very first vision (the vision of the Image) reaches to the end of "the times of the Gentiles," when the Image is smitten, ground to powder, and the Stone by which it is smitten, fills the whole earth. The subsequent visions belong to the same period of which the second chapter treats: they add an abundance of fresh instruction but they do not *chronologically* go beyond the scope of the second chapter. The first vision is as to its subject matter limited, and its instruction, consequently, is narrow as to its scope: whereas, the subsequent visions are wide and diversified in subject, and many of them minute in detail. The visions of the Revelation are arranged on the same principle. The sixth chapter, which we have been considering, extends over the whole period of which the subsequent visions up to the nineteenth chapter treat. Its instruction is general, not specific, extending up to the period of the Lord's return. The subsequent visions do but retrace and lead us to the same end. The words of our Lord in Luke xxi. may be quoted as parallel to the end of the sixth chapter of the Revelation. "And there shall be signs in the sun, and in the moon, and in the stars; and upon the earth distress of nations in perplexity at the roar of the sea and of the surge;* men's hearts failing them for fear, and for looking after those things which are coming on the earth: for the powers of heaven shall be shaken. And then shall they see the Son of Man coming in a cloud with power and great glory." Our Lord describes in simple language, the same event that is signified in the Revelation by a symbolic vision.

But this is not the only passage in the Revelation, which these writers consider to belong to the days of Constantine. One of the objects of the Book of Revelation is to show how differently things are estimated by God in Heaven, and by man upon the earth beneath. Thus golden candlesticks, in the midst of which the Son of Man walketh, are used as the expression of God's heavenly estimate of the honourable position of some in the earth, who, viewed in the light of human judgment, were contemptible and worthless, and counted as the off-scouring of all things. Thus also the kingdoms of the Roman World, when manifested in their utmost glory, and delighted in by men, are symbolised in the Revelation by "a Dragon great and fiery red ($\delta\rho\alpha\kappa\omega\nu$ $\mu\epsilon\gamma\alpha\varsigma$ $\pi\upsilon\rho\rho\sigma\varsigma$), having seven heads and ten horns, and upon his

* See corrected Greek reading.

heads seven diadems;" and the Dragon is said to be "the old serpent, that is called the Devil, and Satan." Such is Heaven's estimate of the kingdoms *most* glorious in the sight of men. And at the same moment when God gives this as His expression of the condition of the Roman Kingdoms, He also signifies His estimate of the only thing which He now sees pure and precious in the earth, viz., His own system of revealed Truth—that system which, however it may be outcast and despised, He will by and by establish by His own mighty power, in the place of supremacy and honour. "A woman clothed with the sun, and the moon under her feet, and upon her head a crown of twelve stars," is the symbol by which God expresses His sense of the excellency of His own Truth—Christ's Truth—now suffering in the earth. The symbol was seen in Heaven, because it is Heaven's, and not Earth's estimate of the value of God's holy system of revealed truth.*

* There are in the Revelation, some symbols which refer to a *corporate* condition of God's people, whilst other symbols refer to them as *individuals*. But there is still another class of symbols which do not refer to *them* at all, but to the Truth which has an abstract existence in the earth, and remains unchanged in its excellency, even when God's people may have failed and fallen from the corporate place of honour which they once held. After the Apostles died, the Church, as to its corporate position in the earth, was not worthy of being represented, and is not represented in Scripture, by any symbol of honour. The "golden candle-

But Mr. Elliott having first determined that the enthronement of Christianity in the days of Constantine, was a work of God's hand in blessing, says that the woman clothed with the sun, etc., represents the Church elevated under Constantine into the Heaven of political power. "The heaven meant," says Mr. Elliott, "is evidently that of political elevation."* (Elliott, Vol. II., p. 772.) And again, "was not the state of things in the Roman Empire" (*i.e.* under Constantine) "one that precisely answered the crisis depicted in the vision?— First, the Christian Church, united as one, and morally bright and beautiful,—abundantly the more

---

sticks," was the last symbol which so represented it. Immediately after the Revelation was given, the Church corporately ceased to answer to that symbol, and lapsed into the world; and it is never again represented by any corporate symbol of honour, until it is made perfect in glory. The condition of the Church at and after the days of Constantine was such that it could not be represented by such a symbol as "a woman clothed with the sun," etc. But God's system of Truth which stands revealed in His word, and is in the earth, cannot lose its excellency. *Christianity* is something that must be distinguished from Christians : *they* may change, *it* cannot.

* It is a favourite thought with Mr. Elliott, that "political elevation" is in this dispensation a blessing. Thus he speaks of the rod given by the angel to John when he was told to measure the sanctuary (Rev. ix.) as representing "the authority of the civil magistrate or ruler" as given to Luther and the Reformers. Thus the very blot and reproach of the Reformation, is looked upon as its honour and blessing.

so from the purifying effect of the late persecution,—appeared before the world ascendant, for the first time, in the political heaven; with the sunshine embracing it of the highest of the three Imperial dignities, and the light and favour of the second also beaming on it: moreover with the chief Bishops resplendent at its head, as a starry coronal." (Elliott, Vol. II., p. 779.)

Surely one might well weep over the delusion which causes any one thus to speak of the corrupted Christianity of the days of Constantine, and to sanctify the evil act of its elevation, by attributing to that act these holy symbols. One born in the days of Constantine, and who lived to witness the effects of his proceedings, says, that the condition of Christianity was such, that one might "weep over it like Jeremiah. To what will these things grow, and at what point will they stop? I fear lest the things now around us should be the smoke of the expected fire, lest Antichrist should come in upon these things, and make our failures and weaknesses the occasion of his own greatness."*

* Gregory Nazianzen, born A.D. 324. His words are just as applicable to the beginning, as to the close of the century in which he lived. Mr. Elliott seems to speak of the Woman appearing in Heaven, as representing the exaltation of the real, as contrasted with the professing Church. But this is refuted by fact. The Christianity exalted in the days of Constantine was Christianity *as a whole;* it was emphatically professing Christianity. Ages before Constantine, the true

And if we turn to the remarks of Mr. Elliott on the twelfth chapter of the Apocalypse, why should he say that "*Heaven*" is not to be understood as meaning "Heaven," but that it indicates the "heaven of political authority and power." If this were so, we might certainly expect to find the symbolic parts of the Revelation habitually employing the word in this meaning. Let us take, then, the first six instances and the last instance in which the word occurs in the symbolic parts of the Revelation.

"After these things I saw, and, behold, a door was opened in *heaven*: and the first voice which I heard was as it were of a trumpet speaking with me, saying, 'Come up hither, and I will show thee things which must be hereafter.'" (iv. 1.)

"Immediately I was in the Spirit: and, behold, a throne was set in *heaven*, and upon the throne one sitting." (iv. 2.)

"And no one was able in *heaven*, nor in earth, * * * to open the roll." (v. 3.)

"And every creature which is in *heaven*, and on the earth, * * * heard I saying, 'Unto Him that sitteth upon the throne,'" etc. (v. 13.)

Church had become an invisible remnant in the midst of a corrupt visible body.

It is very evident that prophetic Scripture can be understood by none, whose consciences are so darkened, that they distinguish not the triumphs of Satan from the works of God.

"And the stars of *heaven* fell unto the earth." (vi. 13.)

"And the *heaven* was separated from its place as a scroll when it rolleth itself together." (vi. 14.)

"He showed me the holy city, Jerusalem, descending out of *heaven* from God." (xxi. 10.)

In these passages (and there are a multitude of like instances), who would think of saying that Heaven meant "*political authority and power*"? If bad political authority and power be meant, then God's throne could not be set therein; if good be intended, then it could not pass away as a scroll under the visitation of the hand of God. Moreover it would be difficult to say in what sense John could be commanded to come up into "political authority and power," or how the Heavenly City could descend from it. If, then, Mr. Elliott be deterred by these difficulties from saying that in these cases Heaven means "political authority and power," why should he deviate from the general meaning of the word throughout the Apocalypse, and arbitrarily assign another signification in the twelfth chapter?

But let us adopt the strange hypothesis, and suppose that in the twelfth chapter of the Revelation, Heaven does mean the Heaven of "political authority and power." How *then* will the interpretation hold? We must believe, if we adopt the system of Mr. Elliott, that Satan after having held political authority and power till the days of Constantine, was at the conversion of Constantine deprived of "political authority and power;" and if so he must have been deprived of it for ever, because it is expressly said that his place was not "found ANY MORE in Heaven." Was then Satan deprived, and that FOR EVER, of all political authority and power when Constantine assumed the profession of Christianity? Did he exercise no "political authority and power" when under Constantine, he strove to persecute Athanasius and to exalt Arius? In the days of Constantius, Constantine's Arian successor, and of Julian the Apostate, and of the Arian Goths, did Satan exercise "*no more*" "political authority and power?" In the days of the Popes, and of the Emperors and Kings of Papal Europe, or in the days of Mahomet and his successors, or in the days of Robespierre or Napoleon, was Satan possessed "*no more*" of political authority and power? Why, even the very chapter which is quoted in support of this extraordinary statement expressly teaches the very reverse; for at the moment when Satan is supposed by Mr. Elliott to be cast down from "political authority and power," a "voice is heard in heaven saying, * * * Woe to the earth and to the sea! because the devil hath come down unto you, having great wrath, knowing that he hath but a short time." In other words, it is a time when his power will be increased and not diminished, and as a consequence, he raises up Antichrist, (it matters not in the present question who Antichrist may be) and Satan gives to Antichrist "his power, and his throne, and great authority, * * and all the earth is astonished after him." All this immediately follows the final ejectment of Satan from the presence of God in heaven, or as Mr. Elliott says, his ejectment from "political power." It would be strange if such undisputed supremacy in the earth, should be the immediate consequence of his loss of all political power.

Mr. Elliott indeed supposes that when it is said that Satan is to be "cast out into the earth" (xii. 9), and again, "Woe to the earth and to the sea! because the devil hath come down unto you, having great wrath, knowing that he hath but a short time,"—Mr. Elliott supposes that "*earth*" does not mean "*earth.*" What it does mean, he does not very explicitly state, but he can be allowed no liberty of choice as to the interpretation. If he insist on saying that "*heaven*" means political power, he is bound to give a corresponding meaning to "*earth.*" If the being in heaven is supposed to signify the possession of political power, the being in the earth, when used in contrast, must signify the non-possession of political power: and inasmuch as those supposed to be exalted into the heaven of political power were *Christians,* those represented by the "earth" must be *Pagans*. In what way then did Satan, when dispossessed (as Mr. Elliott asserts) of political power, attack the non-possessors of political power, *i.e.* the Pagans. In what sense had he great wrath against *them?* Why he must rather have favoured them, for they were his own servants, and how should they be the objects of his rage? Moreover, how does their being the object of Satan's wrath, consist with Mr. Elliott's former interpretation of their being at that moment subjected to the wrath of the Lamb?

But once more; there are few passages in Scripture of more solemn moment than that which, in this chapter, directs our thoughts to the accusations of Satan against us "*before God*," or more literally, "*in the presence of God*," (ἐνώπιον Θεοῦ).* What Satan was in the days of Job and of Jehoshaphat, that he still is—"the Accuser of the brethren," accusing "them before our God day and night." Few verses perhaps are more mysterious than that in Job, which says, "Now there was a day when the sons of God came to present themselves before the Lord, and Satan came also among them;" but it plainly reveals just what this passage in the Revelation declares, that Satan, as the Accuser, is

* "And I heard a loud voice in heaven, saying, 'Now hath come the salvation, and the power, and kingdom of our God, and the authority of His Christ: because the accuser of our brethren hath been cast out, who accused them before our God day and night.'" (Rev. xii. 10.)

admitted even into the presence of God. Mr. Elliott seems to forget the words "*before God*," and treats the passage as if it merely referred to the persecutions conducted by earthly Pagan Princes, acting on the accusation of earthly adversaries instigated by Satan.

But if it were so, there should have been no persecution of any kind, or in any place, since the days of Constantine ; for when Satan is once excluded as the Accuser, his place after that is "found NO MORE in Heaven ;" consequently if he had been excluded in the days of Constantine, all accusation (*i.e.* according to Mr. Elliott, all persecution instigated by him) must from that time have ceased. But persecutions instigated by Satan abounded after the days of Constantine—witness the persecutions under Arianism, and under Julian the Apostate ; and in later days the Inquisition and fires of Popery.

The elevation of Christianity by God into political power, and the exclusion of Satan from it, would certainly have been followed by a peculiar season of tranquillity and peace ; but instead of this, the chapter before us expressly declares that the ejectment of Satan from Heaven is immediately followed by a time of peculiar persecution and distress. "When the dragon saw that he was cast out unto the earth, he persecuted the woman." * If

* That the ejectment of Satan from Heaven is *immediately* followed by the persecution of the woman, is suffi-

then ejectment from Heaven means the loss of political power, how could he who had just *lost* power in the earth, begin, and that successfully, to persecute her who had just *received* power in the earth? And if the imperial favour which Christianity received in the days of Constantine, raised her into "the heaven of political power," surely the subsequent and far more effectual favour of Theodosius should have preserved her there: yet in the time of Constantine, Mr. Elliott speaks of her as being in heaven—in the time of Theodosius, as on the earth. Again, if the Woman when driven into the wilderness, represents "the true, primitive, orthodox Church" (Elliott, vol. II., p. 792), in contrast with the false professing body, does she when seen in Heaven represent only "the true, primitive, orthodox Church?" Was the true Church

---

ciently proved by the word *when* in verse 13. "WHEN the dragon saw that he was cast out unto the earth, he persecuted the woman." Mr. Elliott's system requires that nearly 300 years should elapse between the ejectment of Satan, and the Woman's reaching the wilderness.

Theodosius the Great is supposed by Mr. Elliott to be "THE great eagle" whose wings were given to Christianity, to flee into the wilderness. It is rather strange that an Emperor who far more effectually than any other exalted Christianity, should be the person who helped it into the wilderness. Constantine and Constantius did persecute Athanasius; but Ambrose made Theodosius bow down at his feet.

only exalted into political power? Alas, the Church had not only brought itself into a condition in which no mortal eye could certainly distinguish the true children of the kingdom; but, more than this, the influence of worldly professors had become so predominant long before the days of Constantine, that the history of his reign is little more than a record of their successful agency.

But I forbear to enlarge further on this subject. It may be that Mr. Elliott and Dr. Cumming will live to be convinced that this chapter is as yet unfulfilled. Then they will bitterly regret having written as they have. It cannot be fulfilled, until the *whole* Roman earth—Eastern and Western—has been divided into Ten Kingdoms, answering to the ten horns that were seen on the head of the Dragon. It cannot be fulfilled, until Satan is finally excluded from Heaven, and ceases to be "the Accuser of the brethren before our God." It cannot be fulfilled, until Satan, being confined to earth, enters on the last and shortened period of his power—shortened, because limited to twelve hundred and sixty days. During that time, Christianity will be driven from all the Ten Kingdoms of the Roman World; and blasphemous infidelity, under THE ANTICHRIST, will reign.

§ II.

THE whole prophetic system of Mr. Elliott and Dr. Cumming depends like that of Mr. Fleming, on the assumption that *days*, in the language of Scripture, mean not days, but *years*.

The doctrine that *days* mean years, is a comparatively modern notion. The following quotations sufficiently show the prevailing doctrine on this subject, during the early centuries.

Irenæus, in the second century, says: "When Antichrist shall have ravaged all things in this world, *reigning three years and six months*, and shall have sat in the Temple at Jerusalem, then the Lord shall come from Heaven in clouds, in the glory of the Father, to cast him and those who obey him, into the lake of fire." (Irenæus, Adv. Heres., chap. xxx.)

Hippolytus, in the third century, says, commenting on Daniel: "Time, times, and half a time; by which Daniel means *three years and a half.*"

Cyril of Jerusalem, A.D. 360, says: "Antichrist shall reign *three and a half years* only. I say not this from the Apocryphal writings, but from Daniel, for he says, 'and it shall be given into his hand until a time, &c.;' now a *time* is one *year.*" (Catech. xv.)

Jerome, in the fifth century, says: "*Time* signifies a year: *times* (according to the idiom of the Hebrews, who themselves have a dual number) signify two years; *half a time* six months." Jerome speaks of this as being the doctrine which "*all ecclesiastical writers have delivered.*" (Jerome on Daniel.)

Theodoret, who lived in the middle of the fifth century, says in his commentary on Daniel vii.: "By a time, times, and half a time, the prophet means *three and a half years*, during which that horn which speaketh great things shall prevail."

Bede, who lived in the seventh and eighth centuries, commenting on the twelfth and thirteenth chapters of the Apocalypse, says: "*Time* signifies a year; *times*, two years; *half a time*, six months: for before the three and a half years, he (Antichrist) does not blaspheme openly" (aperto ore).

Quotations to this effect, might be almost indefinitely multiplied—but it is unnecessary. Even Mede allows, that until the twelfth century, all expected an Antichrist "who would last for *three years and six months*" (Triennalem et semestrem expectabant). Indeed Mr. Elliott himself says, "It is, I believe, the fact, that for the *first four centuries*, the days mentioned in Daniel and the Apocalyptic prophecies respecting Antichrist, were interpreted literally as *days*, not as *years*, by the Fathers of the Christian Church. * * * They looked perpetually for *the breaking up of the Roman Empire into ten kingdoms*, as a sign of its near approach: that division being understood by them to mark the time of *Antichrist's* revelation; and, in accordance with the *literal* interpretation of the prophetic days, as the forerunner, *at only three and a half years' interval*, of the coming of the Son of Man. Such was the expectation of Irenæus, Hippolytus, Tertullian, Cyprian, Lactantius, Cyril, Chrysostom, Jerome, and in fine Augustine." (Elliott, Vol. III., p. 966.)*

* Nor are testimonies to this effect confined to writers before the twelfth century.

Sebastian Munster, born in 1489, originally a Franciscan Monk, but eventually one of the Reformers, and said by Dupin to have been one of the ablest of those who embraced the Protestant faith, in his commentary on Daniel vii., writes thus: "God, says the Prophet, will not permit the tribulation from Antichrist and his followers to assail His elect with undue severity; but it shall continue for a *time*, that is a year; and *times*, that is two years; and *half a time*, that is half a year; in other words, it shall not last through a septenary and full period, but through the half of a hebdomad or septenary period. For those days shall be shortened for the elect's sake." (See Critici Sacri *in loco*.) *Clarius* also, a Benedictine of the sixteenth century, and *Vatablus*, Professor of Divinity at Paris in that century, write to the same effect. Their remarks may be seen in the commentaries on Daniel given in the Critici Sacri. See also the commentary of *Zeger* on the Apocalypse given in the same work.

Grotius gives the same interpretation, and quotes in con-

It is of course possible that all these writers may have been wrong. On many most important truths, their teaching was wrong; but in such cases we are able clearly to disprove their statements by the word of God. What ground then is there in Scripture for saying that *days* mean *years?* The arguments used by Mr. Elliott are as follow:

The first is nearly the same as that given by Mr. Fleming. It is argued that because Israel had a

---

firmation the following passage from *Josephus*, who speaking of Antiochus, says, that he despoiled the Temple, and caused the daily sacrifice to cease for *three years and six months.* (Josephus on the Jewish War, Lib. I., cap. 1.) *Clarius* also refers to this: after saying that the saints would suffer under Antichrist for three years and a half—or as stated by John, forty and two months, he adds: "for so long a time the Temple was profaned by Antiochus, who was himself a type of Antichrist."

The passage from Josephus clearly proves that he understood *days* to mean *days.* The following passage shows the doctrine of the Jews on this subject at a later period. *Aben Ezra*, who lived in the *twelfth* century, referring to Saadias, a celebrated Jewish Teacher, says: " Now Saadias expounds correctly and well; know also that in Holy Scripture *days are always days, and never years.* Yet it is possible that the word *days* may mean an entire year, since the repetition of the days produces a return of the year; as when it is said, Ex. xiii. 10, 'from days to days,' that is from year to year; *days* meaning a complete year. *But when the number is stated, as two days, three days, it cannot mean years, but must be days, as it stands.*" (Aben Ezra, as quoted by Maitland. See also Dr. McCaul.)

sabbatical *day*, and also a sabbatical *year*, that this may be an intimation that day is to be understood as meaning year. But surely there can be no better argument to prove that *day* means *day*, and *year* means *year*. Did not Israel understand that every seventh day was a sabbath day, and every seventh year was a sabbatical year? Did they not know that God intended that they should keep the day, and keep the year? They would have been astonished, indeed, if any one had told them that *day* meant *year*, and that *year* meant 360 years; for in that case they ought not to have kept any seventh day at all. "Remember the sabbath *day* to keep it holy" would have meant, "Remember the sabbath *year*." Six years should have intervened between each sabbath day; and six times 360 years between each sabbatical year; and each sabbatical year, when it at last came, should have been 360 years in duration.

Another argument is, that when the spies of Israel had searched the Land for forty *days*, Israel was punished by wandering in the wilderness a corresponding number of *years*. But in this case also "day" means "day," and "year" means "year:" otherwise the spies must have been in the Land for forty years, and the Israelites have been punished for 40 × 360=14,400 years, and therefore be still in the wilderness.

Again, it is argued, that Ezekiel was commanded to lie on his side, and did lie on his side for 390 days, in order typically to bear the iniquity that Israel had been committing for 390 years. In this case also, *days* mean days, and *years* mean years ; otherwise Ezekiel must have lain on his side 390 years, and Israel's iniquity must have extended to 390 × 360=140,400 years.*

Mr. Elliott's next argument is, I believe, peculiar to himself—indeed he allows that it has never before been used in this controversy; and it is this, that there is in Hebrews vii. a text in which daily means not daily, but yearly. The text is this: "Who needeth not daily (καθ' ἡμεραν) as those high priests, to offer up sacrifice, first for His own sins, and then for the people's." On this passage Mr. Elliott observes, on the authority of Macknight, that the High Priest of Israel never offered sacrifice, except once in the year, and therefore daily (καθ' ἡμεραν) must be understood to mean yearly (κατ' ἐνιαυτον).

They who make this very extraordinary statement seem to have forgotten that there is, in the *tenth* of Hebrews, another text in which the word "*daily*" occurs. It is this: "every Priest standeth

* I am aware that the writers who quote these passages, are not accustomed to insist upon the non-literality of the years as they do on that of the days. But they must in consistency do so. If in Daniel the word "time," which all allow to be equivalent to the word "year," is to be understood of a year of years, and not of days; year in the other passages must mean a year of years—and then all the impossibilities mentioned above would follow.

DAILY ministering," &c. If daily means yearly in the *seventh* of Hebrews, we must of course suppose that it means yearly in the *tenth* chapter. Indeed there is no alternative; for seeing that the words "EVERY Priest" must include the High Priest, the same reason which is supposed to necessitate the alteration in the seventh chapter, must necessitate the same alteration in the tenth chapter. If it be true that the High Priest never sacrificed but once in the year, then it cannot be true that "EVERY Priest standeth daily" offering sacrifice. Thus "*daily*" must be altered into "*yearly*" in the tenth of Hebrews also, and we should be obliged to say that there was no "daily sacrifice" offered in the Temple at all.

But it is not so. There is not in the whole Bible an expression of more fixed and unalterable meaning than the words translated "daily" (καθ' ἡμέραν). It is used sixteen times in the New Testament, and never means anything else than "daily." Any one may satisfy himself in a moment by turning to a Greek Concordance.

The notion that the High Priest of Israel only ministered at the altar once in the year, is entirely mistaken, and arises from confounding between the ministration *in white linen garments*, (which was special and peculiar to the Day of Atonement,) and the ministration which devolved on him daily in his *proper priestly garments*, which are fully described in Exodus xxviii. "And for Aaron's sons thou shalt make coats, and thou shalt make for them girdles, and bonnets shalt thou make for them, for glory and for beauty. And thou shalt put them *upon Aaron thy brother*, and his sons with him; and shalt anoint them, and consecrate them, and sanctify them that they may minister unto me in the priest's office. * * * And they shall be upon Aaron, and upon his sons, when they come in unto the tabernacle of the congregation, or when they come near unto the altar to minister in the holy place; that they bear not iniquity, and die. It shall be a statute for ever unto him, and his seed after him." Aaron could never have approached the altar *in these his proper priestly garments* at all, if his ministration had been confined to that of the white linen garments of the Day of Atonement, which was annual: nor could he ever have offered the necessary sin-offerings for himself according to the ordinance in the fourth of Leviticus; and therefore his priesthood must have been defiled; for the fourth of Leviticus enacted that for every sin which the anointed Priest committed in ignorance, he should as soon as it came to his knowledge, immediately offer a sin-offering; otherwise he would bear his iniquity. This *might* be necessary *many* times every day; so untrue is it that he only offered annually. Accordingly, the Epistle to the Hebrews contrasts, not only the annual ministration of the High Priest on the Day of Atonement, but also the *daily* ministrations of

MR. ELLIOTT AND DR. CUMMING CONSIDERED. 237

all priests, with the one offering now once and for ever made, and never repeated. The *daily* as well as yearly sacrifices are contrasted with the one offering—consequently, the word means emphatically *daily* in the two passages quoted.

It may therefore be safely said that not only does Scripture afford no instance of "*days*" meaning "years," but that all the instances quoted in support of this assertion, prove exactly the reverse. They prove that "days" mean "days," and "years" mean "years." It would be strange indeed if it were otherwise; for what possible reason could there be that Scripture should not say days when it means days, and months when it means months, and years when it means years?

Mr. Elliott candidly resigns the argument which is commonly founded on the seventy hebdomads of Daniel, and admits that no proof of days meaning years can be founded on it.* Indeed, even in his mode of using the arguments just recited, it appears to me that there is a degree of hesitation apparent, which seems to intimate, that he is some-

* Mr. Elliott's words are: "I do not rest, in this argument, on the parallel of Daniel's celebrated prophecy of the *seventy weeks:* * * * * and for this reason, because the Hebrew word שָׁבֻעַ, rendered *a week*, has been shown to be a word etymologically of ambiguous meaning, signifying *any septenary*, and applicable to *seven years* as well as *seven days*." (Elliott, Vol. III., p. 962.) For further remarks on this passage of Daniel, see "Prospects of the Ten Kingdoms," p. 217.

what mistrustful of their soundness. Other writers, such for example as Mr. Fleming, boldly refer to these arguments, as if they plainly and satisfactorily proved, that "days" mean "years." Mr. Elliott uses far more cautious language, and speaks, somewhat obscurely, of a "transition from days to years, as if the lesser were a type of the greater." (Elliott, Vol. III., p. 958.) But this very statement does in fact surrender the argument; for the lesser *must absolutely exist*, otherwise it could not be a type of something greater to follow. Therefore, according to this principle, Antichrist must reign for 1260 literal days, in order to become a type of 1260 years, which *must* on this hypothesis follow; for a type precedes an antitype. Thus the fact of Antichrist's literal reign is conceded. We may safely leave the consideration of the 1260 years that are supposed to succeed; for seeing that the millennium immediately follows the reign of Antichrist, there would be no room for the succession of 1260 *years* of woe.

If there were no other argument to be urged against this very extraordinary notion, the partiality of its application would be sufficient for its refutation; for why, when the expressions two days —three days—forty days—three years—seventy years, and the like, occur hundreds of times in Scripture, should the words "days" and "years" be uniformly understood to mean days and years in all these passages, and in one or two other passages, equally definite, be interpreted differently? Why, when it is said of Nebuchadnezzar in his madness, that "seven times" should pass over him, do we say that "seven times" mean *seven years*— and immediately after, when the same word occurs, say that "three and a half *times*" mean 1260 years? Soundness of interpretation requires consistency. If "time, times, and half a time" in the seventh of Daniel mean 1260 years, then seven times in the fourth of Daniel must mean the double of that period, and then we must say that Nebuchadnezzar is yet alive, and his madness continuing still.*

But if we were to allow Mr. Elliott's hypothesis, and suppose that days mean years, even then, his system will not bear the test either of Scripture or of fact, nor is it consistent with itself.

* Mr. Elliott endeavours to obviate the force of this argument, by supposing that Nebuchadnezzar was personally degraded for seven literal years, in order to be a type of the degradation of his Empire for 2520 years. But in this case the literal interpretation as regards Antichrist is granted, and he must reign for three and a half literal years. In this case too, seeing that a type precedes that which it typifies—if the three years and a half of Antichrist's reign were typical, it must precede that which it is supposed to typify, viz., the 1260 years of his mystical reign; just as Nebuchadnezzar's degradation precedes the supposed antitypical period of his Empire's degradation. Thus the notion refutes itself; for Antichrist's literal reign terminates the dispensation of sorrow and of evil.

The events to which the period of 1260 *days*, or as Mr. Elliott says, *years*, belong, are:

The power of the little Horn, as described in Daniel vii.

The sackcloth testimony of the two Witnesses.

The sojourn of the Woman in the Wilderness.

The reign of Antichrist as the sole Head of the Roman World, described in Rev. xiii.

These periods are made parallel or coincident periods in the Scripture, and are rightly so regarded by Mr. Elliott and Dr. Cumming. They begin at the same moment, and they terminate at the same moment. The event that terminates them, is the Advent of the Lord in glory.

Now it is obvious that the period thus marked by 1260, whether days or years, must be a *definite* period. In other words, it cannot extend over a period longer than 1260. There must be a *fixed* moment in the counsels of God (whether we discover it or not) for the commencement of this period; and when it has once commenced, from that moment the numbering of the 1260 must begin. To say that there may be two different periods for its commencement, would be to say, that it was not a *definite* period at all. It would be equivalent to saying that it *did* begin, and yet that it did *not* begin—it would oblige us to say that it *has* ended, and yet has *not* ended.

But Mr. Elliott maintains that there *are* different epochs for the commencement of this period. One he places at the æra of Justinian, A.D. 533 — the other in A.D. 606, when Phocas acknowledged the primacy of the See of Rome; so that, after having commenced in A.D. 533, about seventy years afterwards it commenced again; and consequently, after having ended in A.D. 1793, will, in about fifteen years more from this present period, end again. Therefore, according to the theory we are considering, this most precise and definite period *did*, and that by the appointment of God, commence in A.D. 533, and yet did *not* commence: it did end in 1793, and yet will not end until 1866.*

* The year 1866, so long looked for, has now passed. Would that at its expiration the evil and misleading theories which we are considering, had expired also. But it is far otherwise. Efforts are still being made by astrological calculations, and otherwise, to perpetuate the delusion.

The notion that there may be a *year*-period, and also a *day*-period has been suggested by others besides Mr. Elliott —only *he* considers a *day*-period to come first, and a *year*-period after, whereas *they* imagine 1260 *years* to come first, and 1260 *days* after; and they regard the 1260 years as passed or nearly so.

But such a notion is immediately set aside by the fact that during the last twelve centuries there has been steady progress in the world's prosperity. Civilisation has advanced, and is advancing still—God allows this; even as when He said of Pharaoh, "For this same purpose have I raised thee up, that I might show My power in thee and that My name might be declared in all the earth." God intends that pride should bud and blossom. What likeness can there be between two periods, one of which is a period of advancing

And when we consider the nature of the events that are to close this period of 1260, we cannot but feel more than ever astonished at the supposition of there being a duplicate termination of such a period: for what are the events that are to conclude this period? They are these; the termination of the Woman's sojourn in the Wilderness; the destruction of Antichrist by the brightness of the Lord's appearing—in a word, the abolition of the reign of evil, and the commencement of the reign of righteousness and peace.* Did any of these things occur in 1793? Was the Lord manifested in His glory in 1793, or 1866, or at any past period? Was the reign of Satan *terminated*, in any sense whatever, by the French Revolution? The French Revolution *terminated* no evil. On the contrary, it was a period when the energies of men were set more free than ever to work their own designs; and the result is seen around us on

---

prosperity, and the other a period of judgment, curse, and destruction. The period of the 1260 *days* will be one in which God's hand will be stretched out to smite the strength of man with plagues more terrible than those which He poured upon Egypt of old. What likeness can such a period have to the years of prosperity which precede?

* Nothing can be more express than the words of Daniel: "I beheld, and the same horn made war with the saints, and prevailed against them: UNTIL the Ancient of Days came, and judgment was given to the saints of the High Places;" *i.e.* the reign of the saints immediately follows the fall of Antichrist.

every side, in the continuance of old systems of iniquity, as well as in the development of new. Those periods in the past 1800 years, when God has permitted one form of evil to chasten or to supplant another (and this was the case at the Revolution, when Infidelity warred with Popery) are as different in principle from that great coming hour, when God will interfere, by the mission of His Son, as the workings of Satan are different from the direct agencies of God. In the former case, evil, and that of increased intensity, supplants former evil, and strengthens itself on the ruins of that which it has supplanted: in the latter case, God by His own immediate hand terminates for ever the reign of wickedness, and brings in the kingdom of righteousness and peace. This is the subject of the Revelation. It is not the history of the *progress* of human evil. It is the record of the manner in which God finally smites that evil, and establishes His own glory. Holy angels clothed in white, issuing from the Temple of God, are fit expressions of such agencies of Heaven: but who, unless his mind were under the power of some radical error, could suppose that such hallowed symbols could represent the wickedness and triumphs of prospering Revolutions? Who, unless fatally deceived, could find in the occurrences of 1793, anything similar in principle to that hour, when God will "arise to judgment, to save all the meek of the earth."

But again, let us pass by this error also — let us suppose that the 1260 days, or, as Mr. Elliott says, *years*, have a duplicate commencement and a duplicate end: even then, the system refutes itself. Mr. Elliott and Dr. Cumming suppose this period to have commenced first in A.D. 533, and to have commenced again in A.D. 606—in which case it must have terminated first in 1793, and would again have terminated in 1866. Now since we are told in the eleventh of Revelation that the two Witnesses prophesy in sackcloth throughout the whole of the 1260 days, or, as Mr. Elliott says, *years*, the first termination of their testimony, according to Mr. Elliott, should not be earlier than 1793; and the second termination should not be earlier than 1866. In apparent forgetfulness, however, of the dates they have fixed on for the commencement of the sackcloth testimony, Mr. Elliott and Dr. Cumming go on to say, that the Witnesses died in A.D. 1514, revived in 1517, and ascended in 1555.* Now seeing that the sackcloth testimony cannot be continued after they have died and ascended, (whatever that ascent may mean, for the Scripture expressly says, "When they shall have completed their testimony, the beast shall kill them ")—it fol-

---

* Mr. Elliott supposes the Witnesses to have died, when the Lateran Council was held, and none of the "Bohemian Heretics" appeared to plead before it: their resurrection and ascent is supposed to be at the political establishment of the Reformation. See Elliott, Vol. II., p. 735.

lows, that if they were killed in 1514, the 1260 days, or years, ought *then* to have concluded, whereas, according to Mr. Elliott's previous statement, they cannot conclude earlier than 1793 or 1866.

Another inconsistency is this; if, as Mr. Elliott says, the Witnesses ascended into the "heaven of political power" in 1555, how can the Woman be in the wilderness hundreds of years after they have reached the "heaven of political power?"—for Mr. Elliott supposes the Woman to remain in the wilderness till the middle of the present century. Nor, if the period of Satan's being excluded from heaven was, as Mr. Elliott says, in the days of Constantine, *i.e.* about A.D. 320, must Mr. Elliott be permitted to delay the flight of the Woman into the wilderness until A.D. 600 or 606, *i.e.* nearly three hundred years after. The Scripture expressly says that "*when* the Dragon saw that he was cast unto the earth, he persecuted the Woman," and seeing that the Woman did not, and could not resist, she has no alternative save instant flight and sojourn in the wilderness. It cannot therefore be admitted that there should be three hundred years, or even one year between the exclusion of Satan from Heaven, and her flight.

It is instructive to observe the marvellous inconsistencies into which even thoughtful minds may be led, when they cling to an erroneous system. The truth is, that all these things are future. The

Day of the Lamb's wrath did not come, nor was supposed by men to have come, when Constantine was nominally converted. It will not come until the Lord is revealed in His glory—the armies of heaven following Him. The "Woman clothed with the sun," etc., does not represent the spurious Christianity of the days of Constantine. The vision is future; it describes a scene which immediately precedes the last 1260 days; the Woman being the abstract expression of the essential excellency of Truth apart from the condition of her supposed, or even of her real children. Satan was not excluded from Heaven, neither did his accusations before God cease, when Constantine professed Christianity. That hour of exclusion is still future, and will be immediately followed by the 1260 days of matured Antichristian blasphemy. The two Witnesses did not begin their sackcloth testimony when the Pope was acknowledged either in A.D. 533, or in A.D. 606—nor did they ascend into Heaven when Protestants in 1555 began to bask in the sunshine of princely favour or (to use the words of Mr. Elliott) "were admitted, equally as Romanists, to sit as judges in the Supreme Imperial Chamber." (Elliott, Vol. II., p. 741.) The path of the two Witnesses will be one very different from that which either the Paulicians,* or the Reformers trod. They are

* As regards the Paulicians, whom Mr. Elliott wishes to honour by assigning to them the place of one of the Wit-

reserved for the furnace of Antichristian evil: the place of their testimony will be Jerusalem, "the city where their Lord also was crucified:" the time of their ascent into Heaven will be at the conclusion of the 1260 days of evil, and is immediately followed by "the sovereignty of the world becoming the sovereignty of our Lord and of His Christ." Antichrist did *not* assume the headship of the *whole* Roman world either in A.D. 533 or in A.D. 606; for the Ten Kingdoms over which he is to rule, are as yet undeveloped; neither did God interfere in 1793 to stop the further progress of human evil. On the contrary, the French Revolution was a period when human evil gathered new strength, and rushed forward into fresh channels which Satan opened for its tide.

Whether, therefore, we regard the past, the present, or the future, we have alike reason to sorrow, and to sorrow bitterly over the views of Mr. Elliott and Dr. Cumming. As to the future, their doctrines hide it from us, and throw an impenetrable covering over the very period which it is the object of the Revelation to unveil. But more than this, they tend to hide from us even present facts, such, for example, as that of Satan's accusations against us before God. And, as regards the world's past

---

nesses, he will find, if he more closely examines their history, too much reason to believe that their doctrines were *deeply* tinged with heresy on many most important subjects.

history, its most important periods, such as the exaltation of Christianity in the days of Constantine, the political elevation of Protestantism, and the era of Infidelity's triumph at the French Revolution, are presented in aspects essentially untrue. The secular exaltation of Christianity was not rejoiced over in Heaven, nor recorded in the Revelation in honourable and heavenly symbols; the commingling of Protestants with Papists in legislative councils was not (as the ascent of the Witnesses will be) the result of a command from Heaven, but was a device of the Evil One; nor were the deeds of such men as Voltaire, Robespierre, and Napoleon, similar, either in kind or in effect, to those holy and heavenly agencies of God, which are finally to usher in and establish the kingdom of His Son. We cannot too earnestly deprecate such principles. They render right exposition of Scripture hopeless, and may be said to nullify the very object for which the Revelation is given.

§ III.

IT will, I think, be admitted by all who carefully consider the objections already advanced, that they do entirely invalidate any system of prophetic exposition to which they apply. It is not therefore *necessary* to say anything further. I will, however, briefly consider the remarks of Mr. Elliott on the thirteenth and seventeenth chapters of the Apocalypse.

Mr. Elliott begins by maintaining very properly, that the little Horn of Daniel vii.—the Man of Sin of 2 Thessalonians ii.—the Antichrist of St. John—the Beast of the thirteenth of Revelation—and the Beast of the seventeenth of Revelation—all refer to the same person, or, as Mr. Elliott thinks, a succession of persons.

But Mr. Elliott seems to forget that two chapters which treat of the same person, may treat of him at different periods of his history. Thus for example, if we were describing in symbols the history of the last great Emperor of the French, we should represent him very differently in the early period of his power, whilst the servant of the Revolutionary System through which he rose into greatness—and afterwards, when having destroyed that System.

he became himself supreme. Thus the seventeenth chapter of the Revelation, although it speaks of the same person as the thirteenth, is not coincident with the thirteenth as to time. It treats of the more early history of Antichrist. It supplies his history before the Ten Kings, who finally divide the Roman Empire, give him their authority; and before he and they unite in destroying the harlot.* The thirteenth chapter describes him *after* he has ceased to be the servant of the harlot, and *after* the Ten Kings are crowned—for in the thirteenth chapter, the ten horns which represent them are crowned. The thirteenth chapter as to time, commences when the seventeenth ends.†

The difference of symbol in the seventeenth and thirteenth chapter is very distinctly marked.

In the seventeenth chapter, Antichrist is represented by a Beast with ten horns *uncrowned;* in the thirteenth, by a Beast with ten horns *crowned*. In the seventeenth chapter, he is sustaining a certain wicked system represented by "the Harlot"; in the thirteenth chapter, having destroyed the

---

* The right reading in Rev. xvii. 16 is, καὶ τὰ δέκα κέρατα ἃ εἶδες ΚΑΙ τὸ θηρίον. "The ten horns which thou sawest AND the Beast," etc.

† The most simple method of instruction is, to communicate first the end; afterwards, to narrate the circumstances that lead to that end. This is the method adopted not only in the Revelation, but also in all other books of prophecy. The order is that of narration—not of fulfilment.

Harlot, he is himself supreme. In the seventeenth chapter the Harlot is the object of attraction to the nations; in the thirteenth, he is himself worshipped. The seventeenth chapter is unlimited as to time; the thirteenth, is confined to 1260 days, or, according to Mr. Elliott, *years*.

Mr. Elliott has failed to draw the necessary distinctions between these two chapters. This would be of itself sufficient to invalidate his exposition of them. But when we examine his statement more minutely, we find still further inconsistency. His explanation of the various symbols, is as follows:

The ten-horned Beast acting, as Mr. Elliott says, under his eighth head, represents *the Pope*.

The two-horned Beast of the same chapter, represents *the Papal Clergy*.

The Image of the ten-horned Beast represents *the Papal General Councils*.

The breath given to that Image, represents *the Clergy as voting in those Councils*.

The Harlot represents *the Church of Rome*.

One single question suffices to nullify this extraordinary exposition. If the ten-horned Beast represents the Pope, how is it that the Pope enters on his supremacy by destroying the Harlot, that is, according to Mr. Elliott, the Church of Rome? The words of the Scripture are express: "The ten horns which thou sawest and the Beast, these shall hate the Harlot, and shall make her desolate and naked, and shall eat her flesh, and burn her with fire." If therefore Mr. Elliott's exposition be true, the Pope must have entered on his supreme power as Pope by first destroying the Church over which he rules!

But further, Mr. Elliott and Dr. Cumming are absolutely obliged to alter one of the leading symbols of the chapter, in order to accommodate it to their system. They continually speak of the "*eighth* head" of the Beast, or "the Beast under its eighth head," etc.; and they make this possession of an *eighth* head one of the cardinal points of their system: whereas the symbol both of the seventeenth and of the thirteenth chapter, is a Beast with *seven* heads, and *seven* only. Even if the healing of the wounded head in the thirteenth chapter, made it another head, yet this could have no place in the seventeenth chapter, where no head was wounded, and none healed; nor is a *healed* head *another* head, for if it were *another*, then it could not be a head *healed*. What then becomes of their system? Its pillar is gone as soon as it is proved that the symbolic Beast had *seven* heads only.

It would not be too strong to say, that the interpretation which assigns these two chapters to the history of the Pope or Popery, will not bear the test of one of the criterion points of these chapters. For example, it is said in the seventeenth chapter, that the Ten Kings after receiving power as Kings at the same period as the Beast, concur in giving their power unto the Beast until the words of God are fulfilled. When did this happen to the Pope? When did the Pope receive the homage of *all* the Ten Kingdoms of the Roman World, in its Eastern as well as Western branches, from the Euphrates to Britain? Or, if Mr. Elliott, contrary to Scripture, insists on obliterating the Eastern and most important branch of the Roman Empire, will he say, that the Western branch of the Roman Empire is divided into Ten Kingdoms who have given, and are giving their concurrent homage to the Pope?—for, observe, they continue to give it "until the words of God have been fulfilled;" or to use the words of Daniel, "until the Ancient of Days has come, * * * and the time has come that the saints should possess the kingdom?" According to the system of Mr. Elliott, *all* in these Western countries, whose names are not written in the Book of Life, should worship the Pope. There should be no Mahomedanism in North Western Africa; no worldly Protestantism in England, or France, or Switzerland. For 1260 years past, *all* evil in Western Europe should have been concentrated in Popery alone.

Mr. Elliott's attempt to find various periods for the commencement of the 1260 days (or as he says, years), and thereby to destroy the definiteness of that period, has been already mentioned. If periods marked by dates are not to be regarded as definite, there is certainly nothing in Scripture that can be considered definite! If God be pleased to inform us respecting a coming period of unequalled blasphemy and tribulation, and at the same time to tell us, that it is not to exceed three and a half years—("that for the elect's sake those days shall be shortened")—am I to believe that in order to make this instruction indefinite, He purposely commences this period at two epochs more than fifty years apart (in A.D. 553 under Justinian, and afterwards in A.D. 606 under Phocas*)?—and that the period, having commenced from one or other of these epochs, was to last 1260 *years?* And even if it could be possible for Antichrist to have appeared so many ages ago, and yet for multitudes of real Christians in ancient and in modern times not to have recognised him; if it should be possible for him thus to have been unknown by a large portion of the Church of God during his life, is it possible that he should also have been destroyed, and yet that there should be one real Christian who is ignorant of such an event? The

---

* In page 1009, Vol. III., Mr. Elliott says—"I have noted the epoch of the promulgation of Justinian's Code and Decretal Epistle to the Pope, as that of the commencement (at least primary commencement) of the Papal Beast's 1260 predicted years of prospering." Justinian's era was A.D. 533.

In page 1011, Mr. Elliott says—"It may also be remembered that I noted Phocas's Decree in recognition and confirmation of the Papal supremacy, as perhaps constituting a *secondary* epoch of commencement to the Beast's 1260 years of prospering." The era of Phocas was 606.

MR. ELLIOTT AND DR. CUMMING CONSIDERED. 255

Scripture tells us, that he is to be destroyed by no mere human instrumentality, but by the Epiphany of the Lord in glory: and his destruction is immediately to be followed by the saints possessing the kingdom. Has anything of this kind taken place? If the 1260 years of Antichrist did *in any sense whatever* commence in A.D. 533, they must have ended by his utter destruction, and that, by the Appearing of the Lord in 1793. It is impossible that he should have a primary destruction, and a secondary destruction; or a *gradual* destruction in any sense—for the very point of all the descriptions given of him during the period marked by 1260 is this—that he is at the acme of triumphant power throughout it, which instead of waning, never wanes at all, but is destroyed by the sudden appearing of the Lord from heaven.

There are many like statements on which we might similarly comment. Thus, for example, the Image of Antichrist, which *all in the whole Roman World* will be commanded to worship, is said by Mr. Elliott to mean the General Councils of the Romish Church. But where are these councils now? They ought, if they are the Idol of Antichrist, and if the Pope be the Antichrist, to have been in continuous existence for the past 1260 years, and to have been "*worshipped*" throughout the whole period. But instead of this, nothing has been more rare, nothing of which the Popes have been more jealous, than General Councils: and when they have been assembled, I know not in what sense it could be said that they were worshipped, even by Papists—but certainly they were never worshipped by Jews, Mahomedans, Greeks, or Protestants. If for 1260 years throughout the whole Roman World, men have been, and are, worshipping Papal General Councils, and are punished with death in case of refusing to worship them, none certainly are aware of the fact. We are neither aware of the existence of such an Image, nor of any such commandment to worship, nor of death being inflicted on such as disobey.

But it is needless to multiply examples. It is evident, that whatever system of interpretation be true, that of Mr. Elliott cannot be true; for it is inconsistent with itself, as well as contradicted both by Scripture and by fact.

It should be observed also, that the principle which Mr. Elliott has adopted as his guide in interpreting the symbols of the Revelation, must necessarily and always lead him into error. The symbols of Scripture should be expounded from the Scripture itself, and not by reference to the works of men; but Mr. Elliott, neglecting the guidance afforded by Scripture, seeks for the key of interpretation in some incident of human history, such as the inscription of a coin, or a sculpture upon a ruin, or some practice adopted by men. Thus, the rider on the white horse who goeth forth "conquering and to conquer," must, according to Mr. Elliott, be a Cretan, *because he has a bow, and Cretans had bows* (Elliott, Vol. I., p. 48): and the symbolic horsemen in the ninth chapter must be Turkmans, because prominence is given in the vision to the tails of the symbolic horses, *and the standard of the Turkmans was three horse tails.* (Elliott, Vol. I., p. 335.) Hence the pictures and inscriptions by which Mr. Elliott has illustrated his whole work. We might smile at these things if they were of less solemn moment. Many of Mr. Elliott's analogies are doubtless mere fancies; but when we remember how frequently men, in following their own plans of pride and wickedness, use expressions or illustrations found in Scripture in a sense entirely diverse from that in which God employs them, we may well tremble at a principle which teaches us to seek in the perverted notions and words of men, the key to the interpretation of the word of God. Satan well knows how to deceive by inducing men to use thoughts or expressions borrowed from Scripture, in a way foreign to the intentions of God, and then leading them to take that perverted use as the guide of their interpretation.*

* This is especially to be remembered, in cases where the blasphemous arrogance of Popery has ascribed to itself expressions and symbols borrowed from the Revelation. Instead of such applications being taken (as is not unfrequently done by Mr. Elliott) as guides to the right interpretation, they should be regarded as beacons of warning. Such

Many instances might be given of the manner in which Mr. Elliott forsakes the guidance afforded by Scripture in the interpretation of its symbols. For example, the imagery used in the fifteenth and following chapter of the Revelation, where those who have triumphed over the Beast are seen on a sea of crystal, having the harps of God, and singing the song of Moses, is imagery derived from the events of Israel's history, when, delivered from Pharaoh, they sang their song of triumph, and beheld the waters of the Red Sea separating them for ever from Egypt and its sorrows. The plain allusion to the deliverance from Egypt, might have deterred Mr. Elliott from applying this passage to any event, except the final deliverance of the people of God into their heavenly rest; for when, excepting then, will the antitypical Egypt be visited by the destroying angel; when else will the saints of God, changed and glorified, be found perfect in crystal purity; when else can it be said that the song of Moses could be sung because the Lord had triumphed over His enemies, and delivered His people? But neglecting all these things which might have guided him—forgetting that both the

---

applications are sure to be either perversions or falsifications of the real meaning of the passage. The object of the Church of Rome is to ascribe to the condition of Christianity, as exhibited in herself, the titles and glories that belong to the people of God after they have entered into the heavenly kingdom.

waters of the Red Sea in which Israel were typically buried, as well as the symbolic sea of crystal, alike indicate the possession of *heavenly* purity, and *heavenly* separation by those who antitypically receive the blessings signified, Mr. Elliott seeks for the fulfilment of these things in some place beneath the skies, and in some earthly waters of separation. The first he finds in the island home of the English Protestants; and the latter, in the protecting waters of the British Channel, which is supposed to have kept them from the terrors and evils of Revolutionary France.

Another instance of disregard to the guidance of Scripture, may be found in Mr. Elliott's interpretation of the rider on the white horse in the early part of the sixth chapter. "And I saw, and behold a white horse; and he that sat on him having a bow; and a crown was given unto him: and he went forth conquering, and to conquer." The words "conquering, and to conquer" sufficiently show that there is One only to whom this symbol can apply: others may conquer, but of Him only whose kingdom is "*unto everlasting,*" can it be said, "conquering, *and to conquer.*" It is the description of Him of whom it is written, that "of His kingdom there shall be no end." These words should have been sufficient to prevent Mr. Elliott from interpreting the passage of a Roman Emperor. If instead of searching into human history to establish the fact that one of the Emperors was a Cretan, and therefore an archer, and thereupon inferring that he must be mentioned in this passage—if instead of this Mr. Elliott had searched the Scripture, he would have found how constantly "the bow" is mentioned to indicate that far-searching, rapid, and destructive power which "the King of kings" will exercise, when He is sent "forth conquering, and to conquer." Thus, when His appearing is described in Habakkuk, it is said that "His bow was made quite naked." Israel's address to Him is, "Thine arrows are sharp in the heart of the King's enemies; whereby the people fall under thee." (Ps. xlv. 5.) And again, "Yea, He sent out His arrows, and scattered them; and He shot out lightnings, and discomfited them." (Ps. xviii. 14.) Such is the character of the power symbolised in this passage: but Mr. Elliott, seeing nothing in it that refers to the future, and seeking to history instead of to Scripture for the guide of his interpretation, sinks this description of the final triumphs of the Son of God, into a description of the victories of a Roman Emperor.

The principle therefore that Mr. Elliott has adopted for his guidance in the explication of symbols, would be of itself sufficient to invalidate his interpretations.

§ IV.

IN concluding these remarks, it may be desirable to mention some of the principles which it is needful to observe in interpreting the Revelation.

In the first place, we have to remember, that the Prophecies of the New Testament are supplemental to those of the Old. One great theme of Old Testament prophecy is the condition of the Kingdoms of the Roman World and the condition of Jerusalem at the time immediately preceding the Advent of the Lord in glory. Chapter after chapter describes the dark picture of Israel's evil, when they shall re-gather to their Land in hardened unbelief. Jerusalem is continually mentioned as the great scene of Israel's national blasphemy, and Mount Zion as the place on which the last great Head of the Gentiles glorifies himself against God. It is impossible, therefore, that the Revelation should avowedly treat of the same period, and yet say nothing of Jerusalem, and Antichrist's relation to Jerusalem; for in that case it would bury every leading fact in Old Testament prophecy. The history of Antichristianism given in the New Testament *must* accord with that given in the Old; and if, in our system of exposition, they do *not* accord, it is a sufficient proof that our system must be erroneous. We have also to remember, that the Revelation does not record the steps by which Jerusalem and the nations gradually reach the height of their final greatness. When Israel, after their deliverance from Egypt, slowly rose into power, the steps of their progress are minutely given in the Old Testament. But there is no such history of progress in the Revelation. The prophetic part of the Apocalypse commences when man *has* reached the height of his evil glory; and when the Lord, no longer permitting Satan and circumstances to favour the increase of that glory, interferes to arrest its progress, and finally to crush it by His judgments. Accordingly, the first mention that is made of Antichrist in the Revelation (see chap. xi.) refers to him as already being in the plenitude of his power; and the first allusion to Babylon is the record of its fall. (Rev. xiv. 8.)

We have also carefully to distinguish between the ordinary providential inflictions of God's power, such as the " earthquakes, pestilences, and famines," which in every age occur, and those interventions of His hand which are more direct and special in their character. Thus the famine that drove Abraham into Egypt was an act of God's providence very different from that more wonderful interference of His "outstretched arm," when He sent Moses and Aaron, clothed with miraculous power, to smite the land of Egypt with peculiar plagues: and again, the plagues thus inflicted are to be distinguished from His final and utterly destroying judgments, such for example as that at the Red Sea, which admitted of no repentance, and was inflicted by the direct agency of God's own presence.

The interventions of God in judgment admit therefore of a three-fold classification, first, the ordinary inflictions of His providence; secondly, the special and more direct interventions of His hand, terrible, but still allowing opportunity for repentance; and thirdly, final acts of utterly destroying judgment, allowing no opportunity for repentance. The judgments of which the Revelation treats, fall mainly under the second and third of these classes. They will be the result of an action of God's hand similar to that which He directed against Pharaoh and the land of Egypt; concluding with the visible manifestation of His glory, and the final deliverance of His people— even as when He delivered Israel and overthrew Pharaoh at the Red Sea. Of such judgments and of such intervention we can still say, "the time is at hand." We could not so speak of His ordinary providential judgments: we could not say of *them* that they are at hand, because they ever have been, and are, around us on every side. Ordinary providential judgments are *not* signs of "the end :" but judgments which come under the second class just mentioned (analogous to the plagues sent on Egypt by the hand of Moses and Aaron) will be signs of the end.

The prophetic parts of the Revelation belong to the same period to which those parts of Daniel which are yet unfulfilled belong. Whenever Israel has not only returned to Jerusalem, *but is nationally recognised in Jerusalem*, we may expect that period to begin.

We must also carefully remember that the Revelation, like Daniel, is divided into many separate visions which are independent of each other, and are not chronologically successive. Thus the sixth chapter is a vision complete in itself, tracing briefly and in outline the whole period of God's chastisements, and terminating as soon as it has reached the hour at which the Lord Jesus is to be revealed. Subsequent visions retrace the same period, and fill up the outline. Isaiah and Daniel are also written on this principle.

It must also be observed, that the Revelation (and the same may be said of the rest of Prophetic Scripture), having for one of its primary objects the comfort of those who in the midst of abounding evil fear God and abide in His truth, is accustomed before it details the action of God's hand in judgment, to preface each vision of sorrow with some intimation of the final blessing that is to come, after the judgments have wrought their appointed work. Thus the sixth chapter, before it refers to those dark agencies of judgment which it is its chief object to reveal, speaks of Him who finally shall go forth "conquering, and to conquer, the results of whose triumphs will be to fill the earth with blessing. We find a similar example in the fifteenth and sixteenth chapters, which contain one connected vision. The last of these chapters reveals the hand of God stretched out to smite the glory of the future great Pharaoh of the earth. The scene in that chapter may be said to be laid in Egypt; but the chapter which precedes, places us, so to speak, on the happy side of the Red Sea, where the people of God sing their song of triumph, delivered for ever from the land of their travail, and guided by God's strength unto His holy habitation. The song of triumph is recorded first, before we hear of the dark circumstances out of which it springs. Each separate vision in the Revelation has thus its own preface of blessing.*

---

* The following may be given as a division of the Revelation, on the principles which have been stated, from the sixth to the eighteenth chapters inclusive.

The sixth chapter is a vision complete in itself—its preface of blessing is in the second verse.

The seventh, eighth, and ninth chapters, form one series, and should be read together. The preface of blessing is in the seventh chapter.

The tenth, eleventh, twelfth, and thirteenth chapters, should be read together. The tenth contains the preface of blessing;

These visions of final glory have a needful prominence given to them in the Revelation, because it is a Book especially intended for the comfort of those who live during the failure of the Church's testimony. The Revelation was given to the Church at a time very similar to the period when the Book of Daniel was given to Israel. That

---

the three chapters which follow it, being separate narratives of the same evil period—the eleventh describing Jerusalem during the prophesying of the witnesses—the twelfth describing the sojourn of the woman in the wilderness during the same period, with the addition of some antecedent concurrent circumstances—the thirteenth describing the concurrent period of Antichrist's full dominion.

The fourteenth should be read by itself—its preface of blessing is in the five commencing verses.

The fifteenth and sixteenth chapters form one vision—the preface of blessing is in the fifteenth.

The seventeenth chapter must be read by itself. It describes Babylon *morally;* and shows that Antichrist at first sustains, but afterwards destroys the peculiar system of law, religion, etc., which will give to Babylon its characteristic distinctness. The first verse contains the announcement of destruction, previous to the description of that which was to be destroyed.

The eighteenth chapter describes Babylon physically, as to its outward wealth and greatness. As in the former chapter, the announcement of the destruction is given before the description of that which is destroyed. (For further remarks on these subjects, see " Babylon, its Revival, etc.," also "Thoughts on the Apocalypse," and " Prospects of the Ten Kingdoms," as advertised at end.)

Book was given to Israel, just at the moment when Jerusalem lost the place of honoured separation, which till then it had occupied in the earth: the Revelation was given to the Churches, just at the moment when they too were about to lose their separate place of testimony. Even before the death of the Apostle Paul, Christianity had begun to exhibit symptoms of its approaching fall. The Church was ceasing to be "the pillar and ground of the truth;" and many "seducers" had crept in, who were perverting the pure testimonies of God, and "beguiling unstable souls." The messages sent in the Book of Revelation to the Churches, clearly indicate the progress of this corruption. Although warned, they gave no heed to the warning, and repented not. Their candlesticks therefore were removed: in other words, God refused any longer to own the *corporate* testimony of Christianity, and visible Christianity became, from that time forward, synonymous with visible evil. A remnant has indeed been preserved in the midst of the general ruin, but they have been few and feeble, divided one from the other, and not unfrequently associated with the professing Church in some of its worst corruptions. Unity and strength have been no more restored to the fallen Church than was the presence of the glory of God restored to Israel, after it departed before the desolating power of Babylon and of the Gentiles.

It is *impossible* therefore that the *prophetic* part

of the Revelation should represent the *corporate* condition of Christianity by any symbol of honour; for how could it put light for darkness, or declare that evil was good? When visible Christianity invited the world into its fellowship, and formally identified itself with the nations, it could only be represented by symbols of the same character as those that pertained to those nations—symbols not of honour, but of abomination, such as a "beast dreadful and terrible" (Dan. vii.), or "a dragon great and fiery red, having seven heads and ten horns, and upon his head seven diadems." (Rev. xii. 3.) It could only have been represented by some such symbol as this, because the same principles that regulate the nations have regulated and do regulate the visible Church.

The visions of glory therefore in the Revelation, even when they refer, as in some cases they do, to a relation which the Church is to hold towards the earth, do not belong to any condition which Christianity has ever yet occupied; *nor do they belong to this dispensation at all.* They belong to other ages, and other scenes, when the present season of Satan's triumph shall have ended, and the Day of the Lord have come. They are intended to stand in contrast with every thing that the eye now beholds, and to comfort those who through tribulation are seeking to endure unto the end.

But how should we shock the faith of such—how should we take from them their appointed consolations—if we were to tell them that these visions of glory referred, *not to the future*, but were designed to represent a visible condition of Christianity yet found on earth. Suppose we had said this to any (if such there were) whose eyes were opened in the days of Constantine to discern the real character of that evil hour—suppose we had told them that the visions by which they were sustaining their souls in hope towards the future were not future, but were being fulfilled in the scenes of rising greatness around them: would they not have felt that we were not only depriving them of their hopes, but that we were attempting to hallow, by some of the most sacred passages of God's word, the very evils against which they were maintaining (for God's sake) a tried and sorrowing testimony? Or if we had succeeded in persuading them, we should thus far have become perverters of their souls; we should have turned them back from the truth and testimonies of God, to seek comfort in the false fires of Satan's kindling; until perhaps they would have rejoiced like others in their deceiving brightness, and have said with them, " Aha, we are warm, we have seen the fire." (Is. xliv. 16.)

Mr. Elliott and Dr. Cumming are, I doubt not, sincere in their hatred of Popery and Puseyism. Indeed, who that has learned to appreciate the way of salvation by grace through faith, would not desire to lay down their lives (if need were) in testimony against those deadly systems? But what gave birth to Popery? It was not the growth of a moment. It sprang from earlier principles of evil long fostered. Are we to accept those principles, and sanctify them by the visions of the Revelation?

The principles of evil that are now being systematised and brought into concurrent operation in the Roman World, are not to be attributed to Popery alone. The Latitudinarianism to which even Evangelicalism has in this country yielded, in consenting to recognise men, whose sentiments they know to be essentially and vitally at variance with the Word of God, as teachers who merely belong to "schools of thought" different from their own—all capable, however, of being received, through the elasticity of love, into one and the same circle of service and of communion—Latitudinarianism such as this cannot be less evil than Popery, because it sustains and gives effect to Popery, and to every known form of influential evil besides. It quenches the light of Scripture, and avowedly seeks the basis of its unity in the exclusion of the definiteness of Revealed Truth. To insist on the recognition of such definiteness is now on every hand deemed impracticable. That may be so. But the question still remains, What has caused that impracticability? Is it sin?

If we examine the principles that are now chiefly influential throughout this country—a country that has been so long favoured with the unrestricted circulation of the Holy Scripture, we shall find that they fall under one of two classes—they are either such as lead men into the darkness and death of mere ritual religion, or they are the latitudinarian principles just described. These are at present, the two conflicting systems—both, however, equally characterised by setting aside the supremacy of Holy Scripture.

Of these systems, the latter is that which will finally prevail. Its comprehensiveness (for it is as comprehensive as evil itself) gives it a necessary superiority over a system essentially contracted. A strict system of ritual Christianity is far too narrow for a sphere in which Judaism, Mahomedanism, and every other form of human thought have to be conciliated: it could not adapt itself to such a sphere;—but a latitudinarian system can. The latitudinarian system can honour and use even its rival, if that rival will consent, as finally it will, to take a lower and less exclusive place; and to preserve a partial, though still extensive rule, by resigning its claim to sovereign and universal sway.

When such a system—having wealth, or as men say, "property," for the pillar of its social fabric, and commerce as the channel through which that wealth accrues—has been established in the *Eastern* as well as Western divisions of the Roman World, we shall see before us the sphere of which the Revelation treats—the sphere on which the judgments respecting which it forewarns us are to fall.

Israel will be in Jerusalem, and will supply one chief feature in the scene.

If we read the Revelation and the rest of Prophetic Scripture with this scene before our minds, and remember the principles that have been stated, many a supposed difficulty will vanish, and Prophetic Scripture in all its leading statements will be found as simple as any other part of the word of God. It would then afford substantial subject for our meditation and for our service. We might perhaps be able to meet the revived energies of the nations, by the recovered testimonies of the Holy Apostles and Prophets—recovered out of a darkness in which they have long been buried, as in a grave.*

---

* Many of the foregoing remarks apply equally to writers whose general views are very dissimilar from those of Mr. Elliott and Dr. Cumming. Thus Dr. David Brown, whose book against the pre-millennial Advent of our Lord has obtained such wide circulation, writes as follows:

"It is notorious, too, that a large number of the primitive Christians for three centuries, fell into the same mistake, expecting the struggles in which they were engaged, to issue in the personal appearing of their Lord, and 'the first resurrection' of His martyred witnesses. The militant did indeed become a triumphant Church, but in a very different sense from what was expected. The martyred testimony of Jesus 'lived and reigned,' but the martyrs themselves lived not. The Gospel slew the great red dragon—Paganism was defeated in the high places of the field—Christianity ascended the throne of the Cæsars: that was the reality contemplated.

in the word, but which the enthusiasm of so many had led them to misinterpret."

If our conscience misjudges facts—if we do indeed believe that the great red Dragon was slain, and that the saints arose, according to the twenty-first of Revelation, when Christianity ascended the throne of the Cæsars in the days of Constantine—it is quite certain that the testimonies of the Revelation and all the Prophets will be sealed books to *us*.

CHAPTER XII.

SCRIPTURAL PROOF OF THE DOCTRINE OF THE
FIRST RESURRECTION.

WE must never forget that the bias of the human mind is *against*, not towards, Revealed Truth. It is one of the consequences of the Fall—one of the results of "sin that dwelleth in us," and of the power put forth against us by Satan. In the reception of *natural* truths it is otherwise. There, for the most part, the mind is ready to receive anything that is proposed to it by competent authority. In its own generation it is wise. But in religious truth, men not only manifest an indisposition to receive, but also a disposition to lose that which they have received. What more remarkable example of this than when the Corinthians, even under the Apostle's own eye, began to question the fact of there being any resurrection? "How say some among you that there is no resurrection of the dead?" If, then, among those whom the Apostle himself taught, the fact of there being any resurrection was disputed, can we wonder that in these latter days—days in which the Scripture tells us falsehood would flourish —can we wonder that now the truth of the First Resurrection should be denied? Shall we, because few believe and teach it, say that it is false? Or shall we, remembering the tendency in man's mind to quit its grasp on truth, refuse to be led blindly by the authority of others, and search for ourselves, humbly and diligently, the Scriptures of God?

Resurrection (that is to say, resurrection into that condition of spiritual being, into which all the redeemed will finally be brought) is, at present, seen only in Christ the Lord. He only, as the "First fruits of them that have fallen asleep" (τῶν κεκοιμημένων), has risen into the glory of the new creation of God. Enoch indeed, and Moses, and Elijah exist in Heaven in bodies *miraculously* sustained, and glorious; but they have not yet their spiritual bodies, nor will they have them until the time shall come for all the Church of the first-born ones (τῶν πρωτοτόκων) to rise into their appointed glory— made perfect in the likeness of Christ. "David," says the Apostle, "hath not ascended into the heavens." The spirit of David is indeed with Christ in the paradise of God, there, together with the spirits of all departed saints, comforted and blessed with joys unspeakable; but David, as to the integrity of his person, that is, in *body* as well as soul and spirit, is not there. He is still, as to this, numbered amongst those "that have fallen asleep," whose "first-fruits" Christ is. All, therefore, that Rome teaches respecting the assumption of the Virgin Mary, and the condition of the saints in heaven, is a blasphemous fiction—a lie from the father of lies. Mary and the rest of the saints remain yet, as to their bodies, in the dishonour of death. Their flesh has seen corruption. The worm has fed on them. They wait for the resurrection-morning. To teach otherwise is virtually to teach the heresy of Hymenæus and Philetus, who said "that the resurrection was past already, and overthrew the faith of some."

The resurrection of the Lord Jesus is to us the *manifested* proof that the mighty debt which He undertook to discharge on behalf of His believing people has been fully paid. As the Surety of His people, performing for them what *they* could not perform, and bearing for them what *they* could not bear, Jesus lived and died. But as soon as He was able to say "It is finished," the claim of Divine Justice was satisfied. The debt was fully paid: and this was *proved* when he burst the prison of the grave, where He had lain to fulfil the Scripture, and to show not only the reality of His death, but also that death had no longer power against Him; for it could not prey on Him. His holy body saw no corruption. Death never had any power against Him except as the Surety discharging the debt of His people. As soon, therefore, as that debt had been discharged, Death lost its title against those whose Surety He was: and of this His resurrection afforded the manifested proof. It proved that every claim against them was satisfied for ever.

Moreover, the Lord Jesus rose as "the Firstfruits"—that word being the pledge that all His believing people shall finally be conformed to His heavenly likeness. In Him, too, their new covenant Head, they have life. "Our life is hid with Christ in God." In the Epistle to the Ephesians we read of the exceeding working of God's mighty power "to us-ward who believe, which He wrought in Christ when He raised Him from the dead, and set Him at His own right hand in the heavenly places." This mighty power was towards *us*, because, in raising Christ, it raised Him as our Head. Consequently, believers, though personally on earth, are *representatively* in heaven. In Christ risen they are "quickened;" in Him they are "raised up and made to sit in heavenly places;" all this the result of His resurrection. By resurrection, too, He has entered as our "Forerunner" within the vail, "able to save to the uttermost those that come unto God through Him, seeing He ever liveth to make intercession for them."

There is, however, an hour coming when the life which is now hidden for us with Him in God, will be no longer hidden, because the time of fruition and manifestation will have come. But when, and under what circumstances? We know that the earth is, in the Scripture, promised a period of millennial rest, when "nations shall learn war no more"; when converted Israel shall "blossom, and bud, and fill the face of the world with fruit"; when creation, which now groaneth, shall groan no longer; when the wolf and the lamb shall feed together; when Satan shall be bound; when the Lord shall be King over all the earth, and His name *alone* be exalted; when at last it shall be *truly* said, "O worship the Lord in the beauty of holiness: fear before Him, all the earth. Say among the Gentiles that the Lord reigneth: the world also shall be established that it shall not be moved: He shall judge the peoples righteously. Let the heavens rejoice, and let the earth be glad; let the sea roar, and the fulness thereof. Let the field be joyful, and all that is therein: then shall all the trees of the wood rejoice before the Lord; for He cometh, for He cometh to judge the earth; He shall judge the world with righteousness, and the peoples with His truth." (Psalm xcvi.) Blessed words, that shall be fulfilled in their season! But when the hour of their fulfilment comes, what shall be the condition of those who have fallen asleep in Jesus? Will Abel and Abraham—will the patriarchs, prophets, and apostles, and all who have followed them in the path of suffering for Christ's sake, be shut out from the joy of that morning of life? Will their bodies be still left in the corruption of death? Or, will one of the first acts of their Lord in the day of His manifested glory, be the calling from the grave all who have suffered for His name's sake, that they may share His joy, and administer His power?

One thing certainly is evident, that that day of gladness cannot be until the dark day of evil described in the Epistle to the Thessalonians shall have first run its course. The Apostle there teaches us that the mystery of iniquity then working should continue to work silently, until it should be at last manifested in all its fulness under "the man of sin," "the wicked" or "lawless one." Now, without here inquiring who "the man of sin" is, this, at least, is manifest, that the reign of righteousness and peace *cannot* be coincident with the reign of "the man of sin." Men cannot at the same time obey Christ and Satan. The nations cannot at the same time bow down before Christ and before Antichrist. The mystery of iniquity and its head must cease to be, before truth and righteousness triumph. To what then does this great head of iniquity succumb? By what is he consumed? By "*the brightness of the coming of the Lord*," says the Apostle. Therefore the brightness of the coming of the Lord *precedes* and *introduces* the reign of peace.

But there is another event mentioned in this chapter as taking place at the coming of the Lord. In the first verse the Apostle speaks of "our [*i.e.*, believers] being together gathered unto the Lord." "Now we beseech you, brethren, by the coming of our Lord Jesus Christ, and our gathering together unto Him." If, then, we are to be gathered to Him at His coming (and the fourth chapter of the previous Epistle describes how), it is evident that the resurrection of all who shall have believed in Christ up to the hour of His appearing, shall take place when He comes to destroy "the wicked one," and to introduce the glory of the millennial age.

If, indeed, the history of the earth were to terminate at the coming of the Lord to destroy the wicked one—if there were to be no reign of peace, no period when righteousness shall flourish in the earth, then, doubtless, there would be no second period of resurrection for the saints, and consequently there would be no FIRST resurrection. But if forgiven and converted Israel are to be in the millennial earth "holiness unto the Lord"— if the spared heathen are to receive through Israel the gospel of grace—if Ethiopia shall then, at last, "stretch out her hands unto God," and the millennium be the great harvest-time for gathering souls into the heavenly garner, will they who shall, in the millennial earth, be thus born again through the word of truth, never be brought into resurrection glory? Will they never bear the heavenly likeness of their Redeemer? Will they always remain in an earthy, fallen body? No! this cannot be.

Well, then, if they are to be finally changed and glorified, there must be a *second* period of resurrection unto life: and this, Scripture uniformly teaches.

In the twentieth chapter of the Revelation we find the following passage: "I saw thrones, and they sat upon them, and judgment was given unto them: and I saw the souls of them that were beheaded for the witness of Jesus, and for the word of God, and which had not worshipped the beast, neither his image, neither had received his mark upon their foreheads, or in their hands; and they lived and reigned with Christ a thousand years. But the rest of the dead lived not until the thousand years were finished. This is the first resurrection. Blessed and holy is he that hath part in the first resurrection: on such the second death hath no power, but they shall be priests of God and of Christ, and shall reign with Him a thousand years."

The obvious meaning of this solemn passage is frequently sought to be evaded on the plea that the Revelation is a symbolic book. Now, what does this mean? Does it mean that, because symbols are employed as the medium of instruction in the Revelation, therefore, no *facts* are taught therein? It would seem to be the thought of some that symbols and figurative language can indicate nothing except that which is shadowy and impalpable. It seems to be imagined that plain and intelligible facts can only be taught where no symbols and no figures are: and that the presence of these is a sign that the interpretation may be as loose as the symbols are supposed to be uncertain. Yet does not such a thought impugn the wisdom of God, who has been pleased in the Scriptures so frequently to employ figures and symbols? Shall we say that He intended that the trumpet should give an uncertain sound?

Literal facts may be taught either by simple language, or by figurative language, or by symbolic actions. This is true both in Scripture and in the ordinary intercourse of men.* An Indian may come to me and say in simple language, "I desire to make peace with thee"; or he may use figurative language and say, "Let us bury our weapons beneath the tree of peace"; or he may say nothing; but, silently, in symbolic action, may dig a grave beneath an olive tree, and there deposit his bow and his battle axe. The modes of communication are in the three cases different, but the same substantial literal fact is conveyed by all, viz., his desire to make peace with me. Nor is it otherwise in Scripture. Thus, the future restoration

---

* The only difference is that the symbols of Scripture are sometimes *actually presented to the outward eye*, as when Ezekiel took into his hand two sticks, symbolising respectively Israel and Judah; or the symbols may be presented *in vision*, as in the Apocalypse. This last form of symbolic instruction can be employed only by God.

and union of the twelve tribes of Israel (which when accomplished will be a literal fact) is taught sometimes in simple language, as when we read in Jeremiah l. 4: "In those days, and in that time, saith the Lord, the children of Israel shall come, they and the children of Judah together, going and weeping: they shall go, and seek the Lord their God. They shall ask the way to Zion with their faces thitherward, saying, Come, and let us join ourselves to the Lord in a perpetual covenant that shall not be forgotten." And again in Hosea i. 11: "Then shall the children of Judah and the children of Israel be gathered together, and appoint themselves one head, and they shall come up out of the land: for great shall be the day of Jezreel." In these two passages there is no figurative language, nor any symbol. But in Ezekiel xxxvii. 16, we find the same event taught in symbol. Ezekiel was commanded to take two sticks; to write on one, "For Judah, and for the children of Israel his companions"; and on the other, "For Joseph, the stick of Ephraim, and for all the house of Israel his companions: and join them one to another into one stick; and they shall become one in thine hand." Such was the symbol presented to the eyes of the people, and thus it was explained, "Say unto them, Thus saith the Lord God, Behold, I will take the children of Israel from among the Gentiles, whither they be gone, and will gather them on every side, and bring them into their own land: and I will make them one nation in the land upon the mountains of Israel; and one king shall be king to them all: and they shall be no more two nations, neither shall they be divided into two kingdoms any more at all." Surely, after such an example we cannot say that a plain literal fact cannot be taught by symbol.

But now to return to the passage quoted from the Revelation: the Apostle saw in the vision thrones occupied by persons who sat thereon, and to them "judgment [that is, authority to rule] was given." " I saw thrones, and they sat on them, and judgment was given unto them." He saw, also, those who had suffered for the name of Jesus under Antichrist, in a *disembodied* state (for it is said he saw their "*souls*"); but it is added, he saw them "*live*," that is, become repossessed of their bodies. "They *lived* and reigned with Christ a thousand years. But the rest of the dead lived not (οὐκ ἔζησαν*) until the thousand years were finished."

Let us suppose that to ourselves such a vision had been sent. Suppose that we had seen those whom we had first beheld in a condition of death, LIVE and reign with Christ, should we have needed to be told we had seen a *resurrection?* I think not.

* Such is the right reading in this passage. For further remarks on this subject, see note appended to the present chapter; and also " Prospects of the Ten Kingdoms," page 265.

But what if we had also heard an authoritative voice say, "THIS IS THE FIRST RESURRECTION: blessed and holy is he that hath part in the first resurrection: on such the second death hath no power, but they shall be priests of God and of Christ, and shall reign with Him a thousand years," —could we, after hearing this solemn comment of God, have doubted what the vision was intended to teach? I say "*comment*," for these words were spoken, not to show that "resurrection" was taught in the preceding vision, but to show what resurrection it was, and what was the blessedness and honour pertaining to those who were included therein.*

* The fact is that this passage was not so much intended to teach, as to confirm, the truth of the First Resurrection; seeing that it had been already abundantly taught throughout the Scripture. Even the *Jews* believed that there would be a resurrection of the righteous when their Messiah came to reign: though they believed it after their own carnal manner, making no distinction between the "first man" who was "earthy," and "the second man" who was from Heaven. Thus we read of Rabbi Jeremiah saying, "When you bury me, put shoes on my feet, and give me a staff in my hand, and lay me on one side; that when the Messiah comes, I may be ready." See Lightfoot, Vol. xi. p. 354.

The teaching of the Lord and His Apostles swept away these vain imaginations. They taught that there was indeed to be a resurrection, but that it was to be a resurrection into *unearthly* glory — glory *above* the Heavens, into which "flesh and blood" cannot enter, and that all who do enter, must be freed from their sins by the Messiah's atoning

Observe, also, these words: "They" (the saints) "LIVED and reigned with Christ a thousand years. But the rest of the dead LIVED NOT until the thou-

---

blood and changed into His glorious likeness, as the Head of the new creation of God.

Throughout the Apocalypse the fact of the First Resurrection is *assumed*. It teems with visions of unearthly glory—all of which belong to "the Church of the Firstborn ones," whose *home* during the millennium will be, not earth (though they will reign over it) but Heaven, and the Jerusalem that is above.

The object of the twentieth chapter is not to announce the fact of the First Resurrection, but to declare the time and circumstances that are to be associated with its accomplishment. "I saw thrones, and they sat on them, and judgment was given unto them." That is, John saw the saints enthroned and reigning. Subsequently it is said, "This" [that which thou hast seen] "IS" [i.e. *represents*, it is the symbolic IS] "the First Resurrection." We know from other Scriptures that in it are included all those "who are Christ's at His Coming." They who are specifically mentioned in this passage are those who will encounter the last great outbreak of Satan's rage, when he will come down "having great wrath that he hath but a short time." *Their* conflict will be great; and they will need great consolation. They will read this passage: will see in it the recognition of the value of their sufferings in God's sight, and will rejoice with a joy given by the living power of God's Spirit.

We frequently find in Scripture, blessings pronounced on individual persons or classes of persons, whose devotedness may be made by God the occasion of making known His mercies, but who are not intended to be the sole possessors of these mercies. Thus when the Lord Jesus said to Simon

sand years were finished." Few, I suppose, will
doubt that the last clause of this passage refers to
the general resurrection described at the close of
this chapter. If, then, the word "LIVE," as applied
in the last clause to the wicked dead, means resurrection, can we arbitrarily attach another meaning to the same word in the first clause, and say
it does *not* mean resurrection? No principle that
at present guides us in the interpretation of language would remain if such licence as this were
allowed.

And if a resurrection unto life be not taught in
the passage we have been considering, what evidence is there of there being any resurrection at
all? The same arguments that nullify the force
of the words "LIVE and "RESURRECTION," in
this passage, would nullify these words, or any
like words in any part of Scripture. The reasons
urged against "resurrection" meaning "resurrection" in the central part of this chapter, tell with
equal force against that solemn passage which
describes the last resurrection at the close of this
chapter. The doctrines of a *first* and of the *last*
resurrection stand or fall together. If the first be

Peter—Blessed art thou, Simon Bar-jona . . . thou shalt be
called Peter (πετρος, a rock-man), the blessing was not intended to be confined to him, but to be extended to all
those who should similarly confess Christ as the Rock
(πετρα). All believers are rock-men.

a mere figure of speech, so is the other. Surely we must beware of explaining away on neologian principles, the "true sayings of God." It approaches very nearly to the sin described in the concluding chapter of this book. (See Rev. xxii. 19.)

It has frequently been said of late that the words, "THIS IS THE FIRST RESURRECTION," indicate not a resurrection of *persons* but of *principles*. The principles of the martyrs, which will, as it were, have died out, are, it is said, to be revived in great power at the commencement of the millennium; and this is termed a resurrection.

But does not such an interpretation as this lack the very semblance of plausibility? For, in the first place, how can the *principles* of the martyrs be said to die out, when we know that during the whole antichristian period, on to its very end, they will be manifested in a power unequalled by anything that has been seen among Christ's servants since the Apostles died? And, secondly, how can the *distinctive* principles of the martyrs be found in the millennium, seeing that then the time of the Truth's sufferings will have passed, and, consequently, martyrdom will have ceased to be?

Moreover, is it not said that they who are raised in the first resurrection shall be made "priests and kings unto God and unto Christ"? We can understand this of *persons*, but how could *principles* be made kings and priests unto God? Is it not also said that over those who rise in the first resurrection, the second death shall have no power? Can we speak of *principles* being subject to the second death? "WHOSOEVER was not found written in 'the Book of Life, was cast into the lake of fire." Can *principles* be so spoken of? If, then, they who rise in the last resurrection be *persons*, they who rise in the first resurrection must be *persons* likewise.

But the Revelation is not the only book in which the doctrine of the *first* resurrection is taught. In the first Epistle to the Corinthians, in the chapter which is specially devoted to the subject of resurrection, we find a passage which expressly treats of *the order* of resurrection. " Every one in his own order: Christ the FIRST-fruits; AFTERWARD (ἔπειτα) they that are Christ's at His coming; THEN (or NEXT, εἶτα not τότε) cometh the end." The words translated *afterwards*, and *then* or *next* (ἔπειτα and εἶτα) are what are called "particles of sequence," that is, they indicate succession of events at certain intervals, as when we say, *first, second, third.* Accordingly, in this passage we are taught that the order of resurrection is, *first,* Christ's own resurrection: *secondly,* the resurrection of those who are Christ's at His coming: *thirdly,* the resurrection of those who live between that coming and the time when the last enemy, Death, is destroyed; which event both here and in the Revelation, is said to be at the close of Christ's millennial reign; after which, "He that sitteth upon the throne saith, Behold, I make all things new."

Unless, therefore, we can cancel the words "AFTERWARD" and "NEXT," we must acknowledge that there are two distinct periods in which the saints will rise in life; one when Christ (having, according to Daniel vii., been brought before the Ancient of Days and formally assumed the sovereignty of earth) shall come forth to put down all enemies; the other, when that object shall have been effected by the destruction of Death, the last enemy. Thus both the commencement and the end of the millennial reign of Christ—that is to say, the period when He shall come forth *in order to subjugate* all enemies, and the period when *He shall have subjugated* all enemies, will alike be marked by a putting forth of that glorious power, whereby, as Son of the living God, He will quicken and bring into His own heavenly likeness those of His servants whom death will for a season have grasped. If, then, there were no other passage but this in the Corinthians, we should be constrained to say that there must be a *first* resurrection. I scarcely need point out the accordance of this passage with that in the Revelation.

In the fifth chapter of John, also, we find a passage in which the *first* resurrection is referred to in contradistinction with the *last*. The 25th verse speaks of the first resurrection, and the 29th verse of the general resurrection; but as these verses, in order to their being properly understood, should not be separated from their context, it will be needful to consider the whole passage.

In the 24th verse of the fifth chapter of John we find these words: "Verily, verily, I say unto you, He that heareth My word, and believeth on Him that sent Me, hath everlasting life, and shall not come into judgment (κρισιν), but is passed from death unto life." This verse speaks of the power put forth through the preached gospel—the word that Jesus was then preaching, not in the power of His manifested glory, but in humiliation, as the sower going forth to sow. By that word believers were and are quickened *as to their souls*—"born again," says the Apostle, "not of corruptible seed, but of incorruptible, BY THE WORD OF GOD. And this is the word which by the gospel is preached unto you." But this blessed word, which when received bringeth our souls unto life, reacheth not our bodies. Death still preys upon *them*, and the outward condition of believers contrasts sorrowfully with that of the "new man" within. But it shall not always be so. The quickening power of life that is in the Lord Jesus *has reached* the souls, and *shall reach* the bodies of his saints. "Verily, verily, I say unto you, The hour is coming, and now is, when the dead [οἱ νεκροι, the dead in body] shall hear the VOICE OF THE SON OF GOD; and they that hear shall live." Observe the words, "voice of the Son of God." This is not the word which He preached in humiliation—a word often rejected and despised, that which the Apostle calls "*the weak thing* (το ασθενες) of God, the *foolishness* of preaching." "Voice," is a word denoting authority and manifested power, as when at the grave of Lazarus, "He cried with a loud VOICE, Lazarus, come forth." And when He shall descend from heaven with a shout, it is with the VOICE of the archangel, and with the trump of God. Jesus spake of the hour when this glorious power is to be manifested, as future: "The hour is coming." He said not this of the word of reconciliation which He was preaching; *that* was *present:* but the hour when death shall be effectually triumphed over in the grave is even yet not come. We still wait for it in hope. Nevertheless, although the appointed hour for the full *display* of that power whereby death is triumphed over is yet not come, yet the power whereby this is to be effected was present in the earth when Jesus was present in the earth, He being Himself "the resurrection and the life," One in whom "all fulness dwells." There is no form, no character of glorious power, that shall in the ages to come be displayed in Jesus *glorified*, that was not present in Jesus *rejected*, therefore He adds the words, "*now* is." "The hour is coming, and now is." Of these last words the grave of Lazarus supplied the evidence; the scene there witnessed being the pledge to us that the same mighty voice that called Lazarus from the grave, shall soon be heard by every saint that sleeps. Then death shall cease to sever those that have known and loved one another in Christ, and all will be happy, and all be perfect, because all shall alike bear the image of their Lord. But observe, *all* the dead are not spoken of as raised at that hour. On the contrary, the concluding words of this verse put an express limitation, "THEY THAT HEAR shall live"—as if all the dead would not then hear that voice of power. Nor will they. "The rest of the dead lived not, until the thousand years were finished."*

* It was appointed of God that the Lord Jesus, before He finally closed His earthly ministry, should make it manifest that He was even then possessed of the same power that will be developed in all fulness when He comes to establish His millennial kingdom. Thus while He could say, "The hour IS COMING," He could also add "AND NOW IS."

Three events will especially mark the commencement of His millennial reign. I. The resurrection of the saints. II. The gathering of Israel, who shall say, "Blessed is He that cometh in the name of the Lord." III. The coming in of the Gentiles to own and rejoice in His glory. The typical pledge of these three things was given during the last week of His earthly sojourn. Lazarus, raised from the dead, sat at the table with Him. Here was an earnest of the First Resurrection. "The hour is coming, and now IS." "The next day much people that were come to the feast, when they heard that Jesus was coming to Jerusalem, took branches of palm trees, and went forth to meet Him, and cried,

294   AIDS TO PROPHETIC ENQUIRY.

And see how clearly this intervening period of the Lord's millennial reign is referred to in this passage. *After* He has spoken of the first resur-

---

Hosanna! Blessed is the King of Israel that cometh in the name of the Lord." (John xii. 12.) Here is the typical token and pledge of that which shall be when the vail shall be taken from the heart of Israel, and they shall recognise their Lord. Moreover, "there were certain Greeks among them that came up to worship at the feast. The same came therefore to Philip that was of Bethsaida of Galilee, and desired him, saying, Sir, we would see Jesus." Here was the pledge of the conversion and gathering in of the Gentiles.

The disciples, if they had been at that moment asked to declare their expectations, would doubtless have said, Now the time is come for our Lord to manifest His glory, and for the Kingdom of God to APPEAR. This conviction had been expressed by them only a few days previously. See Luke xix. 11. But they little knew the path which their Lord had first to tread. "I have a baptism to be baptised with, and how am I straitened till it be accomplished." His atoning work was not yet done. If He had been glorified then, He must have been glorified alone. Not only could none have shared His glory: all must have been left to meet the wrath to come, for none as yet had been redeemed. The appeasing blood had not yet been shed. It was necessary that the Lord should by redemption acquire His title to raise His saints, and to restore Israel, and to gather the Gentiles, and free creation from its groan. Therefore, said He, "Except a corn of wheat fall into the ground and die, it abideth alone; but if it die, it bringeth forth much fruit." If He had not died an atoning death, He must indeed have abode alone. None that had ever lived on earth could

DOCTRINE OF THE FIRST RESURRECTION. 295

rection (in verse 25) but *before* He speaks of the general resurrection (in verse 29), He says that the Father had "given Him authority to exercise judgment also" (*i.e.*, kingly rule), "because He is *the Son of Man*." As Son of the living God He quickens; but it is as *Son of Man* that He is to be brought before the Ancient of Days, to be invested with the millennial power of earth. "I saw in the night visions, and, behold, one like *the Son of Man* came with the clouds of heaven, and came to the Ancient of Days, and they brought Him near before Him. And there was given Him dominion, and glory, and a kingdom, that all peoples, nations, and languages, should serve Him; His dominion is an everlasting dominion, which shall not pass away, and His kingdom that which shall not be destroyed." In the eighth Psalm also it is Jesus, as *Son of Man*, under whose feet all things are to be subjected. But His power is not limited to this. He is not only to quicken, first the souls, and then the bodies of His saints, and to give them rule with Himself over the millennial earth. His power is also to reach back over all the past, over all the world of the dead; and He

---

have inherited His glory. The hindrance to all blessing was SIN unatoned for. That hindrance, has, for all God's believing people, been now for ever removed. The great fact of redemption having been accomplished, all the results will, in due season, follow.

shall summon them all to life again, and shall be their Judge. "The hour is coming, in the which ALL that are in the grave shall hear His voice, and shall come forth; they that have done good, unto the resurrection of life; and they that have done evil, unto the resurrection of damnation." This is the last, the general resurrection. All the wicked dead of every dispensation, as well as the countless multitude of those who have lived and died in faith *during the Millennium*, will be there—the righteous rising in the resurrection of life, the others in the resurrection of damnation, to be judged and to be consigned to "the second death." "Whosoever was not found written in the book of life, was cast into the lake of fire."

And observe, what a marked contrast there is between the twenty-fifth and twenty-eighth verses of John v. In the twenty-fifth verse, a *limited* resurrection is spoken of. . "THEY THAT HEAR shall live." But in the twenty-eighth verse it is said, "*All that are in the graves* shall hear His voice." Surely no words can more plainly contrast the two periods of resurrection. Thus, too, the proper succession of these verses is preserved; for their subject is, power appointed by the Father to be exercised by the Son, and its successional development is said to be this: *first*, that which He exercises through His preached word in quickening souls; *secondly*, that which He will, at "the end of the age," exercise in quickening the bodies of His saints; *thirdly*, that which He will exercise during the millennial reign; *fourthly*, that which He will exercise at the close of that reign, when the hour of the last resurrection shall have come.

Nor is the indirect evidence of Scripture less conclusive as to the truth of a FIRST resurrection. How, if there were not a resurrection of special and distinctive privilege, could we explain the words of our Lord, in Luke xx. 35? "They that shall be counted worthy to obtain that age ($a\iota\omega\nu o\varsigma$) and the resurrection from the dead," etc. Here a resurrection is spoken of that cannot be universal, for it will be attained only by those who shall be "counted worthy." It must therefore be distinguishing and peculiar. So also in the Epistle to the Philippians, the Apostle says, "If by any means I might attain unto the resurrection from the dead." He could not thus speak of the general resurrection, inasmuch as all *must* then arise. His words could only apply to a resurrection that is distinguishing and peculiar, which the *first* resurrection is. Observe too, the expression FROM ($\dot{\epsilon}\kappa$, out of, or from among) the dead—$\dot{a}\nu\alpha\sigma\tau\alpha\sigma\iota\varsigma\ \dot{\eta}\ \dot{\epsilon}\kappa\ \nu\epsilon\kappa\rho\hat{\omega}\nu$. It is not the same expression as "resurrection of the dead," $\dot{\eta}\ \dot{a}\nu\alpha\sigma\tau\alpha\sigma\iota\varsigma\ \tau\hat{\omega}\nu\ \nu\epsilon\kappa\rho\hat{\omega}\nu$. Accordingly, it is never used of the general resurrection. Indeed, such words could only be applied to a resurrection that is distinguishing and peculiar. When we say of any that they have been taken out of an assembled multitude, our words imply selection: they imply that those not thus selected are left.*

Nor could the expression "church of the firstborn ones" (ἡ ἐκκλησία τῶν πρωτοτόκων) be explained unless there were a *first* resurrection. Christ is called "The firstborn from the dead," because He has *first* risen into glory. On the same ground a part of the church are called "The church of the *firstborn* ones," because they precede another part of the church, who will in due

---

* The word ἀνάστασις is once used in Scripture in its primary sense of rising again, in the sense in which one rises after a fall. "Behold, this child is set for the fall and rising again (ἀνάστασιν) of many in Israel." In every other place, (and it occurs forty-one times) it is used of resurrection, properly so called, *i.e.* resurrection of the body.

The reason why it is never applied to the quickening of the soul is obvious. In that case, there is a *new creation*, not the rising again of anything that in any sense existed previously. Accordingly, in Ephesians, we read of the "NEW MAN which after God is CREATED in righteousness," etc., and similar expressions are used in the Colossians. Now, "creation" and "resurrection" are contrasted exercises of Divine power, and therefore we never find "resurrection" applied to the quickening of the soul.

Moreover, quickening of the soul must in every case *precede* the resurrection spoken of in the passages above quoted. St. Paul was already quickened when he spoke of pressing on towards the resurrection *from* the dead. The souls of those "beheaded for the witness of Jesus" were quickened before they were beheaded, and consequently before they "lived and reigned."

DOCTRINE OF THE FIRST RESURRECTION. 299

time follow; and form in the new heavens and
new earth (which will be created *after* the millennial heavens and earth have passed away), one
glorified church for ever.

And if we turn to the Old Testament, how conclusive the evidence there! In the last chapter of
Zechariah, for example, we are taught respecting
the coming of the "Day of the Lord," and how it
will usher in the long promised blessing to Jerusalem and to the earth: "In that day Jerusalem
shall be lifted up and inhabited in her place, . . .
and there shall be no more utter destruction, but
Jerusalem shall be safely inhabited." "In that
day the Lord shall be king over ALL THE EARTH;
in that day there shall be one Lord, and His
name one." No one can *truthfully* say that these
things have been fulfilled as yet; no one, unless
he rejects the testimony of the prophet, will deny
that these things will be fulfilled in their season.
But what event is spoken of as *preceding* the period
of blessing to Jerusalem and "all the earth"? The
manifestation of the Lord with all His saints.
That is the preceding event. "His feet (the feet
of the Lord) shall stand in that day upon the
Mount of Olives, ... the Lord my God shall
come, and all the saints with Thee." No one, I
suppose, will affirm that the saints who thus accompany the Lord when He stands upon the Mount
of Olives, will be in a *disembodied* state. If any
should affirm it, they may be referred to the Epistle

to the Thessalonians, where the Apostle teaches us that the saints, in changed and glorified bodies, will be caught up to meet the Lord in the air, and to come with Him. Now, if they meet the Lord in glorified bodies *in the air*, they must be in glorified bodies when His feet stand on the Mount of Olives. If then, as is taught by Zechariah, the saints accompany the Lord when He comes to the Mount of Olives, and if Jerusalem and the earth teem with inhabitants afterwards, there must be two periods of resurrection; for those in Israel and among the nations who are brought into the fold of faith *during the Millennium*, must finally be changed, and bear the likeness of their risen Lord, otherwise, they could not be saved persons. All the saved must finally bear the image of the Heavenly, even as they have borne the image of the earthy.*

Again, in Daniel vii., where the Son of Man is seen to receive "dominion, and glory, and a kingdom, that all peoples, nations, and languages, should serve Him," we are also taught that "the

* During the Millennium, part of the one redeemed family will have their home above the heavens, perfected in the likeness of their risen Lord: the other part, will for a season, dwell in the millennial earth below, in bodies of flesh and blood. Finally, however, they too will be changed and join their brethren who have gone before, and with them form one glorified Church in the New Heavens and New Earth, where this Adamic creation, and all that is in it, shall have passed away for ever.

saints of the high places shall take the kingdom, and possess the kingdom for ever, even for ever and ever": and again, "Judgment was given to the saints of the high places; and the time came that the saints possessed the kingdom." Shall we say that those who will thus reign together with Christ from the "high" or "heavenly places" are disembodied saints? If not, there must be a *first* resurrection.

Again, observe the quotation made by the Apostle in the fifteenth chapter of the first of Corinthians. The Apostle is speaking of the time when "the dead shall be raised incorruptible, and we shall be changed. "For this corruptible must put on incorruption, and this mortal must put on immortality. So when this corruptible shall have put on incorruption, and this mortal shall have put on immortality, THEN SHALL BE BROUGHT TO PASS the saying that is written, Death is swallowed up in victory." These words—"Death shall be swallowed up in victory"—are a quotation from Isaiah xxv. But in what connection do they there occur? They occur in a passage which describes the final forgiveness and blessing of Israel, and the blessing of all nations in connexion with Israel. The passage in Isaiah is as follows: "Then the moon shall be confounded, and the sun ashamed, when the Lord of Hosts shall reign in Mount Zion, and in Jerusalem, and before His ancients gloriously. . . . . . And in this mountain [Zion] shall the Lord of Hosts make unto all people a feast of fat things. .... and He will destroy in this mountain (Zion) the face of the covering cast over all people, and the vail that is spread over all nations. HE WILL SWALLOW UP DEATH IN VICTORY; and the Lord God will wipe away tears from off all faces; and the rebuke of His people shall He take away from off all the earth, for the Lord has spoken it..... In that day shall this song be sung in the land of Judah," etc. If then the words of St. Paul in the Corinthians are to be fulfilled at the time when "the rebuke of Israel is to be taken away from off all the earth," and when they, together with the nations, shall be converted to the Lord, is it not very manifest that the resurrection of the saints precedes the reign of peace? The Apostle distinctly affirms that the words of Isaiah, "Death is *swallowed up in victory*," refer to the resurrection of the saints, and that that resurrection will take place coincidently with the forgiveness and restoration of Israel as described in Isaiah.

In the succeeding chapter too—a chapter devoted to the description of forgiven Israel's confession and praise—we find the following words addressed by the Lord to them: "Thy dead men shall live, My dead body, they shall arise.* Awake and sing, ye that dwell in dust; for thy dew is

---

* Such is the proper rendering of this passage.

DOCTRINE OF THE FIRST RESURRECTION. 303

as the dew of light, and the earth shall cast out the dead." The departed saints of Israel are here spoken of as Christ's *mystical* body, at present dead. The Lord calls them, "My dead body." But in the morning of Israel's joy, they shall arise. They shall then be bound no longer in the bonds of death. Surely, it well becomes the great Head of Israel, in that day of glory, to call those who have mourned with Him over Israel's and the earth's darkness and woe, to rejoice with Him in the day of Israel's and the earth's gladness.

And do we doubt that at present all creation groaneth? "The whole creation," says the Apostle, "groaneth and travaileth in pain together until now." When, then, is this groan to cease? When is creation to be delivered? At "the *manifestation* (or rather '*revelation*,' ἀποκάλυψις) of the sons of God." Such is the statement of the Apostle in the eighth of Romans. Shall we reject it?

Our appointed hope, therefore, is not resurrection only, but resurrection accompanied with certain circumstances of joy and triumph in earth and heaven, which, by God's appointment, throw an added halo of blessing around that hour of life. The coming of the Lord to "reign gloriously," and to fill the earth with the blessings of truth, and righteousness, and peace; the destruction of the great human head of evil—Antichrist; the binding of Satan; the conversion of Israel, and subsequently of the nations; the release of creation from its groan; these, and like events are, *together with* our own resurrection, our appointed objects of expectation and hope. And remember, it is God who appoints these objects, and therefore, it is not for us to substitute for them any self-chosen objects. Truth only sanctifies. "Sanctify them through Thy truth; Thy word is truth." Our expectations are to be guided, not by our own conjectures, but by Scripture. If God had bidden us wait for the general resurrection as our hope, then that would have been the right sanctifying object of expectation. But if He has not commanded us this; if He has commanded us to wait for "the first resurrection," and for the joy of the millennial morning, when the heavens shall rejoice and the earth be glad; if He has been pleased to describe the character of the glory that is then to be ours, and has made the commencement of the day of the joy of Israel, His people, the moment also of *our* being raised into glory, shall we venture to despise these things, and say that they will never be ours, and persist in expecting things that God has told us not to expect—cancelling, as it were, the promises of His holy Word?

I would not desire to speak with undue harshness of those who reject the truth of the first resurrection; but we are bound, for the truth's sake, energetically to resist their statements. It is no light thing to assert, as one well known opponent of these truths is wont to do, that creation no longer groans. What! Have we indeed no ear to hear the groan? Did the Lord Jesus, when He trod this earth, think that creation did not groan? Did the Apostle Paul think so? And surely the condition of creation has not altered since. If creation had ceased to groan, the sons of God would have been "REVEALED" long since in their glory. (Rom. viii.) Are they thus revealed? "Beloved," says the Apostle, "now are we the sons of God, but it hath not yet been manifested (οὔπω ἐφανερώθη) what we shall be." The same principle that would lead us to say that the sons of God have been manifested, and that creation had ceased to groan, would lead us also to say with Hymenæus and Philetus, that the resurrection was past already.

It is no doubt true that many incautious and unscriptural statements have been made by millenarian writers, both in the early centuries and now. But it is not impossible to separate between chaff and wheat. Truth is not to be rejected because its advocates may err.

We may hold the doctrine of the first resurrection and of the millennial reign of Christ without renouncing the truth that Jesus was, and is, and ever shall be, a King. He is also the great Melchisedek *Priest*, and will be for ever. They who are converted in the Millennium will be dependent on His priesthood and intercession, as much as they who are converted now. In the Millen-

X

nium it will still be true, that in Christ there is "neither Jew nor Gentile, male nor female, bond nor free," all being one in Him risen. Nothing that is now *spiritually* true of believers will cease to be true of Christ's people in the Millennium. In this dispensation we forestall the *spiritual* blessings of the Millennium; but, seeing that it is a dispensation of suffering, not of triumph, we have not the *outward* blessings.

But although it is true that Christ sitteth on the throne of God, and exerciseth the power of that throne, having all power in heaven and earth, yet it is also true, that from the days of Daniel until now, the throne of God has *delegated* certain power in the earth to successive Gentile empires, the last of which has not yet run its course. Now, delegated power must of necessity *belong* to him who delegates it; but its immediate administration is in the hands of those to whom it is delegated. This delegated power is to be resumed by the throne of God when the Ancient of Days shall sit, and Christ will enter on the *administration* thereof; and this "taking to Himself His great power," will commence the millennial reign.

Again, even as the *earthly* distinction between male and female, bond and free, subsists now, between those who are nevertheless one in Christ risen (the man having, even in the Church, privileges which the woman has not), so in the earthly arrangements of the Millennium, although governmental precedence will be granted to Israel in the earth ("to thee shall it come, even the first dominion, the kingdom shall come to the daughter of Jerusalem"), yet this will not prevent Jew and Gentile being one in Christ risen.

Nor do we, in saying that Christ and the *risen* saints will reign over Israel and the earth, imply that earth will ever again be their home. *Heaven* will be their home. The saints are themselves called "the saints of the high or heavenly places." We do not doubt that Moses and Elias appeared in glory on the holy mount. They stood on earth, and were seen on earth. Yet earth was not their home. They departed into the cloud of glory above.

Nor is the millennial the final dispensation. It is not "the dispensation of the fulness of times." After the Millennium has passed, He who sitteth on the throne shall say, " Behold, I make all things new"; and then new heavens and a new earth will be made, where no traces of the first Adam will be found; where the flesh will cease to be, and all be formed in suitability to the heavenly glory of the Second Man. The millennial saints will say, even as we now do, "We wait for new heavens and a new earth, wherein dwelleth righteousness." The resurrection of the Church of the FIRST-BORN will be a step onward in the path of blessing, but it will not be the consummation. If these things be remembered, no doctrine connected with the first resurrection and the millennial reign will be found to jar with any part of the faith once delivered to the saints; nor will one part of Scripture be found at variance with another. The Old Testament and the New Testament will be found to bear concurrent witness to the sufferings of Christ and the glory that is to follow, and the suffering saints will not be deprived of their appointed objects of faith and hope.

NOTE.—I have said in the preceding chapter that the statement of the Apostle in 1 Cor. xv. 22, does of itself prove that there are three, and only three, periods of resurrection unto life—the words used (ἔπειτα, *afterwards*, and εἶτα, *then*) being particles of sequence indicating, as used in that passage, that there are two periods of resurrection only subsequent to that of Christ.

This, indeed, has been, by some, denied: and as their argument is plausible, it may be desirable to supply the answer.

It has been argued by some, that although ἔπειτα, *afterwards* and εἶτα, *then*, are particles of sequence, and denote succession as regards the events to which they are respectively applied, yet the enumeration so made is not necessarily an exhaustive one. It does not necessarily include all the classes or individuals that fall within the circle of the things spoken of. Thus (say they) we find in 1 Cor. xv. 5 an enumeration of certain persons who at various times saw the Lord Jesus after His resurrection; and in that enumeration, these particles, εἶτα and ἔπειτα, are employed. "He was seen of Cephas, then (εἶτα) of the twelve; afterwards (ἔπειτα) He was seen of James, then (εἶτα) of all the Apostles," etc.

DOCTRINE OF THE FIRST RESURRECTION. 309

We know, however, from other parts of Scripture, that the enumeration here given is not a complete enumeration. Others, besides those here mentioned, saw the Lord. Hence, it is argued, that although in 1 Cor. xv. 22, etc., there are three periods of resurrection only mentioned, it does not follow that there may not be other periods at which others may rise, who are not included in this enumeration.

This argument seems plausible; but it will not bear examination. No parallelism can be drawn between the two cases. In his enumeration of those who saw the Lord, the Apostle does not first tell us that he is about to speak of ALL those who saw the Lord; but in speaking of the resurrection he expressly declares that he is setting forth the order (ταγμα) of the resurrection of ALL who rise in the resurrection of life, including even Him who has risen as the Firstfruits of those who have fallen asleep. "So in Christ shall ALL (*i.e.* all who are in Him) be made alive. But EVERY ONE in his own order. [Mark the words, ALL and EVERY ONE.] Christ the first fruits; afterward, those who are Christ's at His coming. Then (εἶτα) cometh the end," when as we know from other parts of Scripture, all the saints who have died during the Millennium will rise, and be glorified.

There is, therefore, no parallelism between the two passages. The Apostle does not in the commencement of the chapter profess to speak of ALL those who saw the Lord: but he does profess to speak in the other passage of ALL those who rise in the resurrection of life.

CHAPTER XIII.

REMARKS ON BISHOP WORDSWORTH'S LECTURES ON THE APOCALYPSE.

§ I.

INTRODUCTORY OBSERVATIONS.

IN Genesis xv. 18, we read, "In the same day the Lord made a covenant with Abram, saying, Unto thy seed have I given this land, from the river of Egypt unto the great river, the river Euphrates : the Kenites, and the Kenizzites, and the Kadmonites, and the Hittites, and the Perizzites, and the Rephaims, and the Amorites, and the Canaanites, and the Girgashites, and the Jebusites." The specification of these nations can leave no question as to the locality of the land indicated. It extended "from the river of Egypt to the great river, the river Euphrates." That land God has appointed to be finally the regulating centre of the whole earth. It is Immanuel's land. Its people shall be all righteous. " Thy people shall be all righteous : they shall inherit the land for ever, the branch of my planting, the work of my hands, that I may be glorified." Thence all nations are to be governed. " The law shall go forth of Zion, and the word of the Lord from Jerusalem. . . . . And thou, O tower of the flock, the stronghold of the daughter of Zion, unto thee shall it come, even the first dominion; the kingdom shall come to the daughter of Jerusalem." (Micah iv. 2, 8.) The supremacy of that land is a means which the appointment of God has rendered necessary for the rectification of the earth's government; for until the veil that now rests upon the heart of Israel is removed, and until they are again planted in their own land, the manifested results of redemption, both in heaven above and in earth beneath, are stayed. Until the time come for Israel to be forgiven, Christ will not take to Himself His great power; creation will not be freed from the bondage in which it groans; Satan will not be bound, nor the saints arise in glory. Well, therefore, might David, and Daniel, and all the Prophets, long for the exaltation of Israel and Israel's land, whence all nations are to be controlled for blessing. The expectation of these things was livingly present in the souls of the saints of old. No sorrows, no chastisements, no trials of their faith, no exercise of their patience, destroyed the vividness of their anticipations. And they form the subject of their anticipations still. Their spirits, indeed, rest in blessedness unspeakable above. Heaven is their everlasting *home*. Thence, together with Christ, they are to reign over the earth, and over all things. But they are not yet perfected in glory. Their bodies are still in the dishonour of death. God is not yet glorified in the earth;

Satan is not yet bound; Israel is not yet saved. Consequently, the promises for which the "holy men of old" waited, are still in abeyance. Have we no fellowship with their desires? If not, how can we have fellowship with God who gave them those desires?

The expectation of the disciples of the Lord Jesus touching these things, was no less fervent. They recognised in Jesus of Nazareth the Head of the promised Kingdom. On the last occasion on which they accompanied Him to Jerusalem, we are told that He spake to them a parable "because they drew nigh to Jerusalem, and *because they thought that the kingdom of God should immediately* APPEAR." They remembered, no doubt, the words, "When the Lord shall build up Zion, he shall APPEAR IN HIS GLORY." They knew that they were approaching the Mount of Olives, and they remembered that Zechariah had said, "His feet shall stand in that day upon the Mount of Olives, which is before Jerusalem on the east, and the Mount of Olives shall cleave in the midst thereof . . . . the Lord my God shall come, and all the saints with thee." (See Zechariah xiv.) Remembering these and like words, the disciples thought that their Lord (whose glory they had once momentarily seen on the Mount of Transfiguration) would again suddenly display that glory; destroy His enemies; forgive and spare a remnant in Israel; would exalt Jerusalem, and make it and its people "a praise in the earth." But they erred. Although right in expecting that the kingdom of their Lord would eventually APPEAR, they were utterly mistaken in expecting that He would APPEAR *then*.

Accordingly, the Lord corrected their error. He told them that He was to be as "a nobleman who went into a far country *to receive for himself a kingdom, and to return.*" Until He shall receive that kingdom (and that will not be until, as we are taught in Dan. vii., He shall be brought before the throne of the Ancient of Days, and there invested with the power of earth, when "the Times of the Gentiles" shall be fulfilled), until that time shall come, the people of Christ are to remain as "pilgrims and strangers in the earth," and to share the rejection of their Lord. Their appointment, *now*, is not to reign, but to suffer. The exaltation of their Lord in the Heavens brings no corresponding exaltation to them in the earth. They are left to bear testimony to His truth; and that testimony is by all (except a little flock) to be rejected, or else perverted. The disciples, when their Lord left them, little understood these things, but when the Spirit was sent to them as the Paraclete, they understood perfectly the character of their future course; renounced all *present* expectation of a glorious kingdom, and became content to wait, and to suffer.

But although they had not been brought into a

kingdom of *manifested glory*, they nevertheless recognised themselves as the subjects of a kingdom new and blessed—a kingdom that had laws and principles distinctively its own; having for its King One who is at present hidden in the Heavens; for its laws, the precepts of Christ and His Apostles; and having for its power the SPIRIT sent down from Him who is the Head of the new Creation of God. Israel and the earth were once more to be tried, to prove whether, after they had rejected the testimony to the Lord of glory *humbled*, they would also reject the new testimony to Him *glorified*. That testimony was (as I have already said) still to be given by His servants *in humiliation*. They were still to wear the garb of Nazareth. The exaltation of their Lord in the Heavens brought them into the recipiency of many new *spiritual* blessings, but those blessings did not exalt them in the earth. On the contrary, they brought them into increased trial and reproach. The laws and the principles of the kingdom to which they belonged were utterly opposed to the laws and principles of the world around them. *Their way* was to be narrow: *the way of the world* broad. The relation of Christ's truth to the systems of men, both religious and secular, was to continue that which it had been when the Lord Jesus Himself stood as its witness in the presence of Caiaphas and of Pilate. He was there a witness for the truth: but a despised and rejected witness.

The Apostles well understood these things. They saw the infinite importance of keeping the Church of the present Dispensation in the place of humiliation, and they steadfastly resisted every attempt to raise it into that place of exaltation which converted Israel will nationally hold during the millennial day. They refused to appropriate to themselves the glories reserved for Zion. They would not reign as kings before the time. (See 1 Cor. iii.)

But as soon as the Apostles died, the doctrines and practical habits of the Church changed. For a time the doctrine of a millennial reign lingered. Irenæus and others maintained it, but none taught it scripturally.

The Scripture, in its descriptions of the millennial earth, carefully distinguishes between the condition of the earthly Jerusalem, which, though glorious, will be strictly that of a city inhabited by mortal men in an Adamic earth, and the condition of that New Jerusalem—Jerusalem above the heavens ($'Ιερουσαλεμ$ $ἐπουρανιος$, see Heb. xiii.) into which flesh and blood cannot enter. That celestial city will never enter this *Adamic* earth. During the Millennium, it will be the home of the risen saints —the Church of the First-born ones ($των$ $πρωτοτοκων$) raised in the first resurrection to share the glory of their risen Lord. As glorified, and having spiritual bodies, their condition will of course be altogether contrasted with that of their brethren yet in the flesh, who will, for a season, inhabit the earth below; yet they will be one with them in spirit, and will, together with angels, watch over them, and from time to time, will visit the earth unto that end. Moses and Elijah were seen on the Mount of Transfiguration: they appeared in glory: but earth was not their home. God manifested His glory on Sinai; angels accompanied Him; yet neither God nor the angels resigned Heaven as their home. Israel in the Millennium shall manifestly be brought under the tutelage of heaven, and from their city, the earthly Jerusalem, legislation and instruction shall go forth unto all nations. Wars shall cease; creation be relieved from its groan; the lamb and the lion shall feed together; and millions will, by the preached Gospel, be gathered into the everlasting fold. But the millennial earth, though blessed, will never be brought even into a *paradisiacal*, much less into a heavenly, condition. Death, the last enemy, will not have been destroyed. Corruption, physical and moral, although restrained and counteracted, will still continue to attach to this Adamic earth and all whose home is in it, until He that sitteth on the Throne shall say, "Behold, I make all things new." Then the Adamic earth and heavens shall pass away, and a new earth, and new heavens be created, whose glory shall as much exceed that of the Adamic earth (whether in its paradisiacal, or in its millennial state) as the glory of the Only-begotten of the Father exceeds that of the first Adam, who was earthy. No one, therefore, can teach the doctrine of the Millennium aright, who fails to assign to *the nation* of Israel its distinctive place in the earth's millennial government; and who does not carefully distinguish and contrast the condition of the Church of the first-born ones glorified, with that part of the Church which will be gathered during the Millennium, and have for a season to tarry in the earth. These, as soon as the millennial reign terminates, will join their brethren who have risen in the first resurrection, and will with them, in the new heavens and new earth, form one glorified Church for ever.

Irenæus taught otherwise. He imagined that the earth during the millennial reign of Christ would be restored to its "*primæval*" condition. He held that the righteous dead who would be raised at the commencement of the Millennium, would not then attain their final condition of incorruption, but that they would be brought merely to "a commencement" of incorruption, and become gradually accustomed to comprehend God (capere Deum). "In the times of the resurrection," Irenæus says, "the righteous shall reign in the earth, waxing stronger by the sight of the Lord; and through Him they shall become accustomed to partake in the glory of God the Father, and shall enjoy in the Kingdom intercourse and communion with the holy angels, and union with spiritual beings."

Besides the risen saints, Irenæus taught that there would be others "whom the Lord shall find in the flesh awaiting Him from heaven, and who have suffered tribulation as well as escaped" out of the hand of Antichrist, and that those thus left "shall multiply in the earth, and shall be under the rule of the saints, and shall minister in this Jerusalem;" that is, a restored earthly Jerusalem, which Irenæus imagined would be built after the pattern of Jerusalem above.* Thus all that the Scriptures teach respecting the future of the converted *nation* of Israel during the millennial reign, and all that they teach respecting the unearthly glory of those who rise in the *First* Resurrection was, by Irenæus ignored. The earthly Jerusalem is, according to him, to be the city, not of converted Israel, but of risen saints who are to reign there; and those risen saints are not to have spiritual bodies, but restored natural bodies— bodies adapted to this Adamic earth. If this were so, all that the Apostle Paul teaches in the Corinthians, and elsewhere, respecting the resurrection of the saints must be false, for he teaches that they shall be raised in glory, and bear the likeness of their risen Lord." "This corruptible shall put on incorruption; and this mortal shall put on immortality." We cannot hold the doctrines of Irenæus and those of Scripture together. One or the other we must renounce. Jerome and others felt this.

* See Irenæus, Book v. § 35.

They maintained that the condition of the risen saints was by the statements of millennial writers carnalised. They denounced therefore the doctrine of the Millennium, and pronounced it to be "a fable."

The doctrine of Lactantius (drawn chiefly from the Pseudo-Sybilline writings, which he quotes abundantly) does not materially differ from that of Irenæus. Neither of them, however, fell into the deadly error of teaching that the Church is reigning *now*. On the contrary, they maintained that the present place of God's people is one of suffering and reproach. They remembered the words of John, "I John, your brother, and fellow-partaker in the tribulation, and kingdom, and endurance in Jesus." They received not the falsehoods, soon after enunciated, respecting Satan being bound, and the saints reigning *now*. But when "Christianity had ascended the Throne of the Cæsars," and firmly seated herself there, darkness rapidly advanced. A cursory reference to the writings of Augustine and Jerome will suffice to show its depth. It came in like smoke from the bottomless pit, and has brooded over Christendom ever since.

Augustine once held the doctrine of a millennial reign, but says that he was stumbled by the manner in which it was taught.* If he and

---

* Commenting on the words, "Blessed and holy is he that hath part in the first resurrection," Augustine says, "Those

Jerome had endeavoured to correct the errors of Irenæus and other Millenarian writers, they would have done well. But instead of that they attacked the doctrine of the Millennium *as taught in the Scriptures*, and overthrew it. If Papias and Irenæus had written wildly, *they* wrote more wildly still.

---

who, on the strength of this passage, have suspected that the first resurrection is future and bodily, have been moved, among other things, specially by the number of a thousand years, as if it were a fit thing that the saints should thus enjoy a kind of Sabbath-rest during that period, a holy leisure after the labours of the six thousand years since man was created, and was on account of his great sin dismissed from the blessedness of paradise into the woes of this mortal life, so that thus, as it is written, 'One day is with the Lord as a thousand years, and a thousand years as one day,' there should follow on the completion of six thousand years, as of six days, a kind of seventh-day Sabbath in the succeeding thousand years ; and that it is for this purpose the saints rise, viz., to celebrate this Sabbath. *And this opinion would not be objectionable, if it were believed that the joys of the saints in that Sabbath shall be spiritual, and consequent on the presence of God; for I myself too, once held this opinion.* But, as they assert that those who then rise again shall enjoy the leisure of immoderate carnal banquets, furnished with an amount of meat and drink such as not only to shock the feeling of the temperate, but even to surpass the measure of credulity itself, such assertions can be believed only by the carnal. They who do believe them are called by the spiritual Chiliasts, which we may literally reproduce by the name Millenarians." (Augustine, "The City of God," Book xx. § 7.)

Their perversion of the Scripture was worse, and the results more disastrous.

The following extracts will give some notion of *Augustine's* teaching. Commenting on the words, " And he cast him [Satan] into the abyss," Augustine says, " By the *abyss* [bottomless pit] is meant the countless multitude of the wicked, whose hearts are unfathomably deep in malignity against the Church of God; not that the devil was not there before, but he is said to be cast in thither, because, when prevented from harming believers, he takes more complete possession of the ungodly."

"And shut him up, and set a seal upon him, that he should deceive the nations no more till the thousand years should be fulfilled." "'Shut him up,'—*i.e.* prohibited him from going out, from doing what was forbidden. And the addition of 'set a seal upon him' seems to me to mean that it was designed to keep it a secret who belonged to the devil's party, and who did not.... By the chain and prison-house of this interdict the devil is prohibited and restrained from seducing those nations which belonged to Christ, but which he formerly seduced or held in subjection.... The devil, then, is bound and shut up in the abyss that he may not seduce the nations from which the Church is gathered, and which he formerly seduced before the Church existed. For it is not said 'that he should not seduce any man,' but 'that he should not seduce the nations'— meaning, no doubt, those among

Y

which the Church exists—'till the thousand years should be fulfilled,'—*i.e.*, either what remains of the sixth day which consists of a thousand years, or all the years which are to elapse till the end of the world." Augustine, "City of God," lib. xx. § 7.

I could quote much more, but I prefer to consign further extracts to the Appendix. If there be one thing emphatically declared in Scripture and momentarily evidenced by experience, it is the fact that Satan is at present *unbound*. He is a roaring lion, going about, seeking whom he may devour. Nor are any nations on earth more fatally and thoroughly blinded and deceived by him than the nations out of which the Church has been taken. They are, indeed, the very nations which are giving birth to Antichristianism, and over whom Antichrist is to reign. How then shall we characterise teaching such as this of Augustine? What words would be strong enough to condemn it? Yet this is the kind of teaching by which Christendom has been for seventeen centuries deceived.

As soon as the Teachers of the Church resolved to maintain that the present is the *final* Dispensation appointed to this Adamic earth, they of course found it necessary to affirm that *every* millennial promise is being accomplished *now*. "The times of refreshing from the presence of the Lord" *have* (according to them) already come; Satan *has been* bound. Israel *is* blossoming and budding, and filling the face of the world with fruit. Creation *no longer* groans; the First Resurrection *has* come. The saints *are* reigning with Christ. Nations learn war no more.* All these, and like things (it is said) have been accomplished, or are being accomplised *now*. At a moment when tears and tribulation peculiarly abound, we are asked to believe that tears have been wiped away from off all faces, and that sorrow and sighing have fled away. The doctrine of Transubstantiation is not more repugnant to Scripture, to fact, and to reason, than are the statements by which the Teachers of post-apostolic Christendom have sought to convince us that the present age of darkness and sorrow is the new age of light, joy and glory. According to Jerome, only Jews and Judaisers believe that the

* Thus Jerome's comment on the words, "*they shall beat their swords into ploughshares*," is, "All desire after war shall be changed into the pursuit of peace: and instead of discord there shall be, throughout the whole world, concord; and lances shall be changed into pruning hooks; so that abandoning the fierce passions of war, men should devote themselves to the culture of the earth and reap abundant harvests, which indeed may also be understood spiritually when all the hardness of our heart is broken by the yoke of Christ and the thorns of vice are eradicated. The words 'nation shall not rise up against nation' were, according to Jerome, accomplished when a census of the Roman Empire was taken in the time of Augustus Cæsar. If that were so, the nations should never have learned or practised war from that day to the present; for it is said, "*Neither shall they learn war ANY MORE.*" (See Jerome *in locum*.)

prediction of Isaiah respecting the wolf and the lamb dwelling together will receive a literal accomplishment. These predictions were fulfilled when Paul who was of the tribe of *Benjamin*—a tribe that ravined as a *wolf*, was baptised by Ananias, and consorted with Peter, to whom it was said, "Feed my *lambs.*" Paul, therefore, was the wolf; Ananias the lamb; and Peter the shepherd!! We read also in Isaiah that the leopard shall lie down with the kid; and the calf and the young lion and the fatling together, and a little child shall lead them. "Daily," says Jerome, "are these words fulfilled, for in the Church we behold the rich and the poor, kings and men of low degree, associated under the teaching and governance of children, that is to say, of Apostles and Apostolic men who are rude in speech, though not in knowledge." The simple words found in Scripture is the straw which the lion eats. The infant also, that is to say, one who is in malice a child, puts his hand upon the hole of the asp when he drives off from the bodies of men, the devils by whom they have been besieged. "The weaned child, that is, one who is able to dispense with milk, and to take strong meat, puts his hand on the cockatrice den, when he puts it into some dwelling-place of Satan and thence drags him out." The holy mountain, in which nothing is to hurt or destroy, is not Mount Zion (as Jews and Judaising Christians say) but *the Church*, whence all the lions, leopards, and creatures

of earth are to be overspread with Truth, even as
the waters of the sea cover that which is beneath
them. Such is the teaching of Jerome and his
School—a School in which Origenism, Mysticism,
and Neologism may flourish; but Truth perishes.
It would seem as if Jerome had no ear to hear
the groan of creation; no heart to pity the throes
of suffering humanity. He stigmatises as Judaisers
all who regard relief from pain and other like
sorrows as a blessing worthy to be spoken of by
the Prophets. He avows that he agrees with the
Stoics in saying that there is nothing good but
virtue, nothing evil but vice. Health of body and
plenty are not to be looked on as blessings. They
are to be numbered "among *things indifferent.*"
When the Lord Jesus in answer to the enquiry of
John, cured many of their infirmities and plagues,
and of evil spirits, and gave sight to many that
were blind, Jerome (had he been present) must have
said, The removal of things like these, "things
indifferent," is no sufficient proof of the presence
of the Kingdom of God. His hard-hearted
Stoicism might prepare for the use of the racks,
and torments of the Inquisition, but certainly it
would lead no heart to welcome the peace and
grace of the Millennial reign. If it be asked how
it was that the Church became so blinded as to
accept falsehoods so palpable, the answer is, their
consciences were darkened, and their hearts were
perverted. They wished to reign as kings. They

coveted the glories of Zion, and longed for exaltation. "Kings," said they, "shall be our nursing fathers, and their queens our nursing mothers." They could only assume the place they lusted after, by consenting to say that the Millennium had already come. They hesitated not to affirm this, and so to adopt a heresy more deadly and more widely destructive than that of Hymenæus and Philetus.

Any one who reads the writings of Eusebius (the flatterer of Constantine), or the sermons and dogmatic books of the Papists (such for example as the Catechism of Pope Pius IV.), will see how all those parts of Scripture which speak of the glories of Zion and of Jerusalem in "the age to come," are grasped by them, perverted, and made the foundation of their claims to supremacy in the earth. They have virtually pronounced the supremacy of Satan to be the supremacy of God, and have confounded a period during which Scripture declares that the diadems of earth are on the head of the Dragon, to be identical with that in which the sovereignty of the world shall become the sovereignty of our Lord and of His Christ. What darkness can be deeper than this? What delusion more deadly?*

* Augustine, commenting on the words of Rev. xx. "And I saw thrones and them that sat upon them," etc., says, "It is not to be supposed that this relates to the last judgment, but to the seats of the rulers, and to the rulers themselves

It might have been hoped that the light of the Reformation would have dispelled this darkness. But it did not. They who were in derision termed "Fifth Monarchy Men," seem to have given a kind of testimony to the truth; but it was partial, connected in many cases with serious doctrinal errors, and in no case rose higher than the level of Irenæus and Lactantius. Their system, accordingly, (ridiculed rather than argued against by Lightfoot and others,) perished under the weight of its own errors. Lightfoot for the most part followed the lines laid down by Augustine and Jerome. Like them he maintained that the First Resurrection is past, and that Satan was bound when the Gospel began to be preached. Truths which Jerome and Augustine spared, Lightfoot cancelled. For example, the final division of the Roman Empire into Ten Kingdoms and the reign of Antichrist in Jerusalem, they taught; but both these great truths Lightfoot ignores. Lightfoot's writings have materially influenced the views of Protestants, both within and

---

by whom the Church is governed. And no better interpretation of judgment being given can be produced, than that which we have in the words, What ye bind on earth shall be bound in Heaven, and what ye loose on earth, shall be loosed in Heaven." Augustine, "City of God," lib. xx. § 9. Lightfoot follows the same line of thought, but finds the evidence of the saints' reign in the secular rather than in the ecclesiastical magistracy. Such in their view is the First Resurrection!!!

without the Church of England, down to the present time. The lectures of the Bishop of Lincoln, which we are about to consider, are an example. The writings of Augustine and Lightfoot have materially guided his thoughts, as may be seen from his references.*

* Lightfoot's views may be judged of from the following extracts: "Observe, by the way, the phrase, 'He shuts him [the Devil] up'; that is, restrains him from seducing and deceiving people. It is Hell and prison to the devils not to be doing mischief. They sleep not, rest not, if they do not evil. It is a torment to them, if they may not be sinning.

"Well, how doth Christ 'bind Satan' that he do not deceive? By sending the Gospel to undeceive them. So that this is the victory of Christ against the Devil. The very telling of His death and merits is that that overcomes the Devil: the very word of His death and resurrection is that that overthrows the Devil and his power. So is 1 Cor. vi. 3 to be taken, 'Know ye not that we shall judge angels?'" (Lightfoot, Vol. VI., p. 15.)

"The Old Serpent had deceived the nations (τὰ ἔθνη) the heathen, for above two thousand years, with idolatry, false miracles, false oracles, and with all blindness of superstition. Now Christ, sending the Gospel by His Apostles and ministers among the heathen or Gentiles, bound the Devil and imprisoned him, curbed his power and delusion, that he should not deceive the world in manner as he had done; but the world now becomes Christian, and heathenism is done away; and this is then called 'the first resurrection'— viz., the resurrection of the dead heathen . . . . Well thus 'the Devil is bound a thousand years' during which time the gospel runs through the world, and prevails and makes it Christian. At verse 7 the Devil is let loose again, and, by Popery, he makes the world as blind, de-

The fact is (and it is a truth that cannot be too solemnly laid to heart) that the teaching of Gentile Christendom has systematically set itself against the doctrine of the eleventh of Romans, and has virtually cancelled that chapter, though it is the great *prophetic* chapter of the Apostolic Epistles. It speaks specifically to Gentile Christendom, and

---

luded, heathenish, as it had been in the worst times under heathenism. And it were no hard thing out of history to show that Rome Papal did equal, nay exceed, Rome heathen in all blindness, wickedness, cruelty, uncleanness, and in all manner of abomination. But an hour, a day, a week, would not serve the turn to describe that full parallel. The text gives a full summary of all, though in few words, when it tells us, that the Dragon, the Devil, 'gave his power, seat, and authority' to Rome: and it hath, and doth, and will act in that spirit, while it is Rome; and can any thing but mischief be expected from such a spirit." Lightfoot, Vol. VII., p. 157.

Perhaps, there never was a period more marked with flagitiousness and misery, physical and moral, than the years that intervened between the Apostolic era and the establishment of Popery—yet during that period, according to Lightfoot, Satan was bound, and ceased to deceive the nations !!

During this period also he says, the saints were reigning and judging angels—the greatness of the kingdom under the whole Heaven having been given to the people of the saints of the Most High, when Christians were established as magistrates and rulers in the Roman Empire. According to Lightfoot, Christ has come and set up His kingdom through the world, and put it into the hands of Christians, "that they should rule and judge in the world, as those heathen monarchies had done all the time before." (P. 260.)

declares that if it continue not in God's goodness, its doom (like that of Israel before) will be *abscission.* "To thee goodness, if thou continue in His goodness, otherwise thou also shall be cut off"—no welcome announcement to those who persistently ascribe to themselves an indefectible standing in the earth, believe themselves to be the Zion of God that standeth fast for ever, and claim that to them all nations should bow. Growth, progress, and triumph, they regard as the very law of their being; and seeing that God's faithfulness is pledged to them, how is it possible that they should fail? "We," say they, "are His people, His chosen, who are to blossom, and bud, and fill the face of the world with fruit? Have we not blossomed? Look at the Palaces, the Castles, the Temples, the Crowns and Banners of Christendom. Have we not marked upon them all the token of the Cross? Are we not advancing still?" So speaks the cankered Gentile olive branch, "proud in its own conceits," utterly unconscious of its approaching doom, and finding in the very things that evidence its shame, the tokens by which it persuades itself that it has eminently continued in God's goodness. And as to Israel, if Christendom condescends to think of them at all, it is to contradict all that the Scripture teaches respecting their wondrous future. The leaders of Christendom, "boasting themselves against the natural branches," affirm that Israel having forfeited their national blessings, those blessings have been made over to *them;* and that no blessing, no salvation, is to come to Israel, save that which now, from time to time, reaches individual Israelites, when, owning Christ, they are graffed into the Gentile olive branch. Is this that which the eleventh of Romans teaches respecting the future of Israel? Does it not positively tell us that Israel, *as a distinct branch*, are *collectively as a nation* to be graffed back into their own olive tree, and that their restoration is to be as "life from the dead" (not, indeed, to Gentile Christendom, for the time of its abscission will have come) but to the spared heathen nations, and to the earth generally, whose day of jubilee will have come?" But all this the Teachers of Christendom (speaking generally) have, for the last sixteen centuries and more, persistently rejected. That their views are unchanged, and their course unaltered, is too clearly proved by the Lectures we are about to consider. They were preached by the present Bishop of Lincoln before the University of Cambridge in 1848. They have the sanction, therefore, of high authority, and have been extensively disseminated.

§ II.

IN his commencing Lecture Bishop Wordsworth acknowledges that "there is scarcely any book in the Bible whose genuineness and authenticity were more strongly attested on its first appearance, than the Apocalypse." (Page 3.) Papias, Irenæus, Justin Martyr, and Melito, who lived in the next age after it was written, positively attest its genuineness.

Nevertheless, no long time after, that is to say "in the third and fourth centuries of the Christian era, many private persons, and even Churches, especially in Oriental Christendom, questioned its canonicity." (Page 3.) And why? Because it taught the doctrine of a millennial reign!

On this I would only observe that these "many private persons and even Churches" must have been thoroughly convinced (and that after careful examination) that the doctrine of a millennial reign was unquestionably taught in the Apocalypse: otherwise they would not have placed themselves in the invidious position of rejecting that holy book, and trampling under foot all the evidence upon which its authority is demonstrated. There must, however, have been great self-will, great insubjection to the authority of God in these persons; and it might have been expected that Bishop Wordsworth would have recognised their sin, and visited it with his severest censure; for surely it is no little sin to reject a book of Scripture because it teaches a doctrine that is to us distasteful. Yet Bishop Wordsworth censures them not. On the contrary, he seems rather to sanction than condemn their course, for he says, "If it [the doctrine of a Millennium] were a true doctrine, then Christ's promises to the Church would have failed." (Page 39.) And again, "If it [the said doctrine] *could be proved from the Apocalypse, the Apocalypse would not be Scripture.*" (Page 27.) If this be right, I do not see how Neologianism is wrong. If instead of resting on the irrefragable *evidence* by which God has sustained the authority of His Word, they to whom the Scripture is sent are to become its judges, and to receive what it pleases them to receive, and to reject what it pleases them to reject, there would in that case, be no recognition of the authority of God, but simply an assertion of the right of self-government grounded on the possession of a verifying faculty in ourselves. Whether the possession of such a faculty be claimed for individuals or for the Church collectively, matters not. In any case the claim is a false one. God, after having established by adequate evidence the authority of His written Word, commits it, it is true, to the *guardianship* of His people; but He makes them not its judges. To reject the Scripture is to reject God. Here then, is a point of unspeakable importance, on which we must be hopelessly at variance with Bishop Wordsworth. The gulf which separates us on this point cannot be bridged over. No wonder that on other subjects there should be no fellowship of thought.

Bishop Wordsworth stigmatises the doctrine of a millennial reign as one that sprang out of ignorant Jewish tendencies of thought that were found occasionally even in Christ's disciples. No doubt wrong tendencies of thought were from time to time manifested by the disciples; but the Lord did not sanction those wrong tendencies. On the contrary, He uniformly corrected them. But did He correct them in the two cases which Bishop Wordsworth cites as instances of erroneous expectation? "The literal interpretation," says Bishop Wordsworth, "of what is spoken figuratively in Holy Scripture, especially in the Prophets, has been one of the most prolific sources of error. . . . . . It suggested the question of St. John himself, and of his brother St. James, 'Grant, Lord, that we may sit, the one on Thy right hand, the other on Thy left, in Thy kingdom.'" (Page 7.) Such was the petition of John and James: but did the Lord Jesus rebuke them, or say that their request was founded on ignorance and error? Did He say that there should be no such kingdom, or no such sitting on His right hand, and on His left in that kingdom? He said exactly the reverse. His answer was: "To sit on My right hand and on My left, is not mine to give, but it shall be given to them for whom it is prepared." (Mark x. 27.) The Lord Jesus therefore admitted the rightness of their anticipation. Was *He* wrong in doing this? Is He to be charged with error?

The other example given by Bishop Wordsworth is the question of the disciples recorded in Acts i. 6. "Lord, wilt Thou at this time restore again the kingdom to Israel?" In replying the Lord does not say, "Ye err, not knowing the Scripture nor the ways of God: there will be no kingdom restored to Israel." On the contrary, His answer distinctly implies that there would be such a restoration. "It is not for you to know the times or the seasons, which the Father hath put in His own power." These words distinctly sanction the expectation of the Apostles, and virtually declare that the kingdom would come at the time and season appointed of the Father. How, indeed, could He who came not to destroy the Law and the Prophets, but to fulfil, say otherwise. Do we not find these words in the Prophet Micah, "Unto thee shall it come, even the first dominion; *the kingdom shall come to the daughter of Jerusalem.*" (Micah iv. 8.) And is not Jerusalem addressed thus in Isaiah; "Arise, shine; for thy light is come, and the glory of the Lord is risen upon thee. For, behold the darkness shall cover the earth, and gross darkness the peoples: but the Lord shall arise upon thee, and His glory shall be seen upon thee. . . . . The nation and kingdom that will not serve thee shall perish; yea, those nations shall be utterly wasted." (Isaiah lx.) All this is utterly untrue now, but it shall be fulfilled in every jot and tittle "when the Lord of hosts shall reign in Mount Zion, and in Jerusalem, and before His ancients gloriously." (Isaiah xxiv. 23.) It is not the twentieth of Revelation only that teaches the doctrine of a millennial reign. The Old Testament teems with testimony thereunto. If then, on the two occasions referred to, we concur with Bishop Wordsworth in condemning the questions asked by the disciples, our condemnation cannot be restricted to them only: it must be extended also to that Holy One who sanctioned those questions. Do we purpose to do this? Shall we impute to the Lord Jesus ignorance, error, and Jewish prejudice?

That the doctrine of the Millennium as taught by Papias, Irenæus, Lactantius and others is thoroughly unscriptural, is true. It is not wonderful, therefore, that such men as Dionysius and Jerome found little difficulty in overthrowing it. But Millenarianism as *taught in Scripture*, and Millenarianism as *taught by men*, are two very different things. I have already referred to the errors of some of the early Fathers on this subject. I will only now observe, that if Bishop Wordsworth had confined himself to the pointing out these errors, he might have done good service to the Truth. But he has not done this: he has presented for our acceptance another system—a system which is virtually that of Jerome and Augustine, and Lightfoot, and other assailants of Millennial Truth. When we examine the system which has thus been constructed on the ruins of patristic Millenarianism, we are compelled to say that that which has been so constructed is incomparably worse than that which has been overthrown. Irenæus erred in his interpretations; but he did not wrest and torture the Scripture as his opponents do.

Whether or not I am justified in saying this, the following remarks will show.

§ III.

THE Apocalypse commences thus: "The Revelation of Jesus Christ, which God gave unto Him, to show unto His servants *the things that must shortly come to pass.*" Again, when the sealed book or roll was about to be given from the Throne, it was said to John, "Come up hither, and I will show thee things that must be HEREAFTER."

With these words before us can we doubt that every event symbolically denoted in the Apocalypse, from the sixth chapter onward, was, at the time when the Vision was given, FUTURE? That the Apocalypse is emphatically a *prophetic* book, is a truth so patent, as, seemingly, to require no proof.

Bishop Wordsworth, however, refuses to acknowledge this. He affirms that the greater part of the twentieth chapter of Revelation (that is to say, all that part which describes the binding of Satan, and the first resurrection, and the millennial reign) *was not when written intended to be a prophecy of the future, but was given as a record of the past.* He thinks that it was accomplished at the Cross, many years before the Revelation was written. We may marvel that Bishop Wordsworth should have come to such a conclusion, but we must remember that his antipathy to the doctrine of a millennial reign is so great that he has said, "If it [the doctrine of the Millennium] could be proved from the Apocalypse, the Apocalypse would not be Scripture." (Page 27.)

Bishop Wordsworth evidently shrinks from avowedly repudiating the inspiration of the Apocalypse: yet indirectly he does this in saying that the twentieth chapter is not *prophetic;* for all that comes within the scope of the roll given from the Throne (and the twentieth chapter does come within that scope) is in the Apocalypse itself declared to be *prophetic.* In saying this, has the Apocalypse said what is false? If so, it cannot be inspired.

Bishop Wordsworth, however, persists in saying that the twentieth chapter (or, at any rate, that part of it which treats of the binding of Satan and the millennial reign) is *not* prophetic. He avers (in this following Lightfoot) that it treats of something that had been accomplished before the Apocalypse was written. His doctrine is, that Satan was bound and "consigned to his dark prison-house," when the Lord Jesus suffered on the Cross.* This he again and again affirms: yet there is really not one of his many arguments that will for one moment bear the test either of Scrip-

---

* Bishop Wordsworth does not here *appear* to deny that Satan is described in the twentieth chapter as personally imprisoned, nor does he say that the bottomless pit is not really the bottomless pit. Subsequently, it will be found that he admits neither of these things.

ture, or of fact. He draws, for example, an utterly false analogy between the condition of the Lord Jesus in death, and a supposed state into which he imagines Satan to have been brought by that death. As Christ died, was buried, placed in the grave, and a seal set on the door of the sepulchre, so Bishop Wordsworth imagines that Satan, when Christ died, was, as one brought under the power of death, buried in the bottomless pit, and a seal set upon him there. But all this is fiction. Satan was an *agent*, not a *sufferer* at the Cross. He there "*bruised*" the heel of the Holy One, brought as the Substitute of His people under the power of judicial death. It is true, indeed, that the *kingdom* of Satan received at the Cross a blow that will ultimately be found to be its death-blow; but Satan himself was not smitten at the Cross. No stroke of the Divine hand fell on him personally. He remained what he was before, "the god of this age"—"the prince of the power of the air"—"the spirit that now worketh in the children of disobedience"—"the Accuser of the brethren," and he, and the spirits that are under him, are declared to be the κοσμοκρατορες, "the world-wide rulers" of the darkness of this present age.* The

* It is quite true that the *title* to bruise the head of Satan was acquired by the great Surety at the Cross: but the *possession* of a title, and its *enforcement*, are two different things. The punishment adjudged to Satan has not yet been inflicted. By the adjudication of God, founded on the

Apostle Peter did not think that Satan had been bound when, writing to the saints, he said, "Be sober, be vigilant: because your adversary the devil, as a roaring lion, walketh about, seeking whom he may devour." (1 Peter v. 8.) Nor did Paul think him to be bound when he spoke of himself as being "hindered" by Satan; and when, seeing other of Christ's people similarly tried and buffeted, he said, "The God of peace *shall bruise* Satan under your feet *shortly*." It may be truly said that there is not one page of Scripture, or one page of human history that does not attest, not only the perpetuation, but the increase, of Satan's power during the last eighteen hundred years. "The annals of Christendom," said Lord Bolingbroke, "have been the annals of Hell." Can this be denied? Yet, notwithstanding the evidence both of Scripture and of fact, Bishop Wordsworth maintains that when Christ died, Satan was "*chained in his dark prison-house*," and that "*the Cross restored the wandering nations to life, health, peace, and joy.*" (Page 52.)

Has Bishop Wordsworth forgotten that the Lord Jesus Himself especially treated of this very subject? Did He say that after His departure from the earth "the nations would be restored to life,

---

title acquired by the Lord Jesus at the Cross, Satan is sentenced to a deprivation of all the power he acquired through the sin of Adam, and is sentenced to everlasting torment: but the sentence pronounced on him *is not yet carried into effect*, nor will be, until Christ returns in glory.

health, peace, and joy"? He says the very reverse. "Nation," said He, "shall rise against nation, and kingdom against kingdom: and there shall be famines, and pestilences, and earthquakes, in divers places. All these are the beginning of sorrows. Then shall they deliver you up to be afflicted, and shall kill you: and *ye shall be hated of all nations* for my name's sake. And then shall many be offended, and shall betray one another, and shall hate one another. And many false prophets shall rise, and shall deceive many. And because iniquity shall abound, the love of the greater part [των πολλων] shall wax cold." (Matt. xxiv. 7–12.) A very different picture this from that which Bishop Wordsworth has drawn. We read nothing here of "life, health, peace, and joy" being "restored to the wandering nations." The teaching of Bishop Wordsworth stands *in direct* contrariety to that of our Lord. Which shall we follow?

And when we ask where and in what Bishop Wordsworth finds this happy change among the nations, his answer is, "Christ at His coming chained Satan who had chained the nations. Bel boweth down, and Nebo stoopeth, before whom myriads had laid prostrate. The idols whose altars had reeked with human blood were cast to the moles and to the bats. The Oracles are dumb. The pagan temples become Christian churches. Basilicas are now Cathedrals. The Cross, once the scandal of the world, floats on the banners of armies, and is set on the diadems of kings." (Page 50.) Such is Bishop Wordsworth's reply. Does he really ask us to believe that in the early centuries of the Christian era, when the lords of the Gentiles sought to sanctify their ambition, rapine, licentiousness, lust, and revelry by assuming the Cross as their emblem, does he really ask us to believe that we are to find in that hypocritical assumption the evidence of Satan being bound, and the reign of Truth having come? Was there any one essential *doctrine* of Popery that was not germinant, and more than germinant, in the days of Constantine, and before? Would that Bishop Wordsworth would remember the warning words of Hegesippus.* And as to Basilicas being turned into Cathedrals, is there in all the past history of the world any

* Hegesippus, A.D. 157, wrote thus :—"The Church continued until then [the commencement of the second century] as a virgin pure and incorrupt; whilst if there were any that attempted to pervert the sound doctrine of the saving Gospel, they were yet skulking in dark retreats ; but when the sacred company of the Apostles came to the end of their course, and the generation of those who had been privileged to hear their inspired wisdom had passed away, then the conspiracy of godless error commenced by the fraud and delusion of false teachers who, *as there were none of the Apostles left* [the deadly doctrine of Apostolic succession was not accepted by Hegesippus], thenceforth attempted without shame to put forward their knowledge falsely so called, and to bring it into antagonism with the preaching of the Truth." Hegesippus as quoted by Eusebius, Eccl. Hist. lib. iii. § 33.

thing more dark—nay, I will say, *so* dark, as the history of the Cathedrals of Christendom? Of what, at this present moment, are St. Paul's and Westminster Abbey the symbols in the estimate of God?* They are standing monuments not only of the past, but of the present, triumphs of Satan over God's people and God's truth. At the present moment, Sacerdotalism reigns in the one; Latitudinarian Babylonianism in the other. Is it in such things as these that we are to find the evidence that Satan has been placed in the bottomless pit, and that the Cross brought "to the wandering nations life, health, peace, and joy"? "If the light that is in thee be darkness, how great is that darkness." When our consciences judge wrongly of facts, our hearts will not be open to the instruction of the Word of God. When our conscience bears false witness, we may be very sure that our understandings will not be allowed to apprehend, nor our heart permitted to feel, the appeals of Truth.

One of the authorities cited by Bishop Wordsworth is a clause from the second verse of the second chapter of Ephesians. I will cite the whole passage, and distinguish by italics the clause quoted by Dr. Wordsworth. "Wherein in time past ye walked according to the course of this world, ac-

* I do not mention St. Peter's at Rome, for Bishop Wordsworth's system requires him to say that Satan was unbound during the days of mediæval Popery.

cording to *the prince of the power of the air*, the spirit that now worketh in the children of disobedience." Does this passage teach us that Satan is "chained in his dark prison-house"? It teaches the very reverse. It teaches us that he is "the prince of the power of the air" (as he ever has been since the Fall), and that he "NOW worketh in all the children of disobedience." He could not be this, or do this, if he were shut up in the bottomless pit. The Bishop proceeds to say, "When He [Christ] ascended up on high, He led captivity captive, and gave gifts to men; enabling them also to wrestle against principalities and powers, against the rulers of the darkness of this world, against spiritual wickedness in high places, and He bruised Satan under our feet. THUS Christ bound Satan, and cast him into the bottomless pit, and shut him up, and set a seal upon him in order that he should deceive the nations no more." (Page 49.) These words, "*Thus* Christ bound Satan," etc., are Bishop Wordsworth's own words; yet they evidently nullify all that he has before said respecting Satan having been placed in the bottomless pit, *by the work of the Lord Jesus on the Cross;* for if that work had placed Satan in the pit, he could not have been placed there by the wrestling of the saints afterwards. These words also show that Bishop Wordsworth does not recognise "the bottomless pit" as the name of an *actual locality;* for we cannot suppose that he imagines the wrestling of the saints to be an agency capable of binding Satan, and placing him actually in the bottomless pit. He looks on the binding of Satan, and the shutting him up in the bottomless pit, as figures of speech, that indicate nothing more than a curtailment or restriction of his power, which restriction is continued if the saints wrestle with him successfully, but removed if they fail in so wrestling. It would be necessary to accept in its entirety Cardinal Newman's system of "*non-natural*" interpretation before we could acquiesce in comments such as these. Moreover, a host of difficulties would still remain unsolved. If the conflicts of God's people with Satan and their victories over him could avail to bind him, he must have been bound *before* the Cross, for from the time of Abel downwards the people of God *wrestled*. The eleventh of Hebrews details many victories of faith. Moreover, if victories of faith could bind him, he must certainly have been bound during the Apostolic period, for then, most unquestionably, he was wrestled with and overcome by very many. But that he was not bound *then* is evident, for he is emphatically declared to have been "a roaring lion, going about, seeking whom he might devour;" and the Apostle Paul speaks of himself as withstood and hindered, by the devices of Satan. Are we then to say that the post-apostolic period supplied more successful wrestlers than that of the Apostles themselves? Were the Apostles' children, and their successors giants? And are the evidences of these mighty triumphs to be found in the transformation of Basilicas into Cathedrals, and the emblazonment of the Cross upon the banner of kings? It must certainly be admitted that the Apostolic age did not supply *such* evidences.

Yet, although Bishop Wordsworth is quite ready to accept the change of Basilicas into Cathedrals, and suchlike transformations, as supplying adequate proof that Satan has been bound and shut up in the bottomless pit, it is very evident that he is not a little mistrustful of his own conclusions. He sees that it may be urged that even nations that have had Cathedrals in their cities, and the Cross on their banners, have not apparently been free from the deceiving power of Satan; and even if it were otherwise, if such nations were free from the power of Satan, their number is comparatively small; whereas the Scripture speaks of "*the nations*," that is, *all* the nations being delivered from the presence and power of Satan during the whole period of his being bound, whether that period be brief or long. Moreover, nations that once had the Cross on their banners, have substituted for it the Crescent, and the eagle, and various other emblems. On Bishop Wordsworth's own principles, this evidences the presence, not the absence, of Satan's power. Can these difficulties be met? Bishop Wordsworth endeavours to meet them by taking a further step in the path of self-contradiction.

He meets them by denying *virtually* that Satan has been bound at all. I will quote his own words. "It is *not* said that Christ rendered it *impossible* for the Devil to deceive; but that He did *His* part, *in order that* he should NOT DECEIVE." (Page 50.) And again: "If Satan still has power in the world (*as doubtless he has**), this, let us remember, is due to man, and not to God. It is because men sleep, when God bids them to watch; therefore the enemy comes. But the Devil is chained to all who do not loose him by their own sin." (Page 49.)

The Bishop's theory, then, is this. Christ (not indeed without the co-operation of His saints whom He strengthened to wrestle) chained and placed Satan in the bottomless pit as respected *them;* but as respects "the nations" generally, seeing that they did not so wrestle, Satan was not chained, nor is he chained to any except those who are careful not to loose him by their own sin. Obviously, then, the casting Satan into the bottomless pit means nothing more than a check or restriction placed upon his power: and that restriction affects only a few; and even as respects those few, the restriction is contingent: it depends on their being careful not to loose him again by their own sin. Consequently Satan is bound and not bound, imprisoned and not imprisoned, at the same moment. Bishop Wordsworth says, "It is not said [in Scripture] that Christ rendered it

---

* These italics are mine.

impossible for the Devil to deceive." Is this so? I find in the Revelation these words—"I saw an angel come down from Heaven having the key of the bottomless pit and a great chain in his hand. And he laid hold on the Dragon, that old serpent, which is the Devil, and Satan, and bound him a thousand years." Observe, it is not said in this passage that the angel laid hold of Satan *in order that he might* bind him. It is said that he laid hold on him *and bound him*, and that for a thousand years. The binding is spoken of as an accomplished fact, *contingent upon nothing*, and irreversible until the thousand years were finished. How, then, can Bishop Wordsworth say, "It is not said [in Scripture] that Christ rendered it impossible for the Devil to deceive"? Could he deceive when shut up? The shutting up *is not spoken of as a contingency. It is spoken of as an accomplished fact.*

Bishop Wordsworth says that there is no reason why we should take a thousand years definitely to mean a thousand years. We reply, that there *is* a reason; for we cannot suppose that God would say what He does not mean. Why should He say "a thousand years" if He does not mean "a thousand years"? If He had wished to mark the period as indefinite, would He not have avoided the use of numerals, and used some general expression? This He does immediately afterwards when the period during which Satan is loosed is mentioned. Of *that* period He does not define the length, but says, "he must be loosed for a short season." The length of Satan's *imprisonment* God has been pleased to define: the length of the subsequent *release* He has not been pleased to define. Had He not some purpose in doing this? Shall we countervail that purpose, and alter His words?*

Nothing can be more true than the rule laid down by the Hebrew writers that whenever periods of time are denoted in Scripture by affixed numerals, the period is literally that which the affixed numeral or numerals denote. It may suit men, for the sake of their own theories, to deny this; yet so it is. But suppose it were otherwise? Suppose that a thousand years means ten thousand years, or ten years, or ten months, or anything else that any one may choose to conjecture, how is Bishop Wordsworth's position helped thereby? It must still be admitted that Scripture does speak of Satan being bound for *some* definite period, however long, or however short that period may be.

---

* We are not at liberty, whenever we please, to divest numerals of their definitiveness, because they may sometimes be used symbolically. "Three" not unfrequently denotes *repetition;* and "seven," *completeness;* but we do not, on that account, say that they are never to be understood definitely. We believe that our Lord rose from the grave on the *third* day; and that Nebuchadnezzar was punished for *seven* years. We continually hear men say, "I have told you a thousand times"; but we do not understand that they use "thousand" with like indefiniteness when they are avowedly making a calculation, or historically stating the length of a period.

Suppose it be a thousand hours? During those thousand hours it would be impossible for Satan to deceive the nations. What becomes of Bishop Wordsworth's assertion? He says, "It is not said in Scripture that Christ rendered it impossible for the Devil to deceive." In that case the Scripture can nowhere have spoken of Satan being bound. Will Bishop Wordsworth affirm this? Or will he say that whether bound, or whether loosed, Satan could in either case equally deceive? It is not very apparent which of these alternatives Bishop Wordsworth would choose. In either case Scripture is contradicted; its light is quenched; and we are left in darkness. The further we advance, the greater becomes the perplexity.

NOTE.—With reference to the words "that he might not deceive"—*ἵνα μὴ πλανήσῃ*, Dr. Wordsworth observes, "It is not said that Christ rendered it impossible for the Devil to deceive, but that He did His part in order that he might not deceive."

It is quite true that the words, *ἵνα* and *ἵνα μή*, may be used to indicate a proposed end which God may allow to be frustrated; or they may on the other hand express a fixed purpose that cannot be frustrated. *The context will, in each case, determine this question.* John v. 34 supplies an example of a proposed end frustrated—"These things I speak that ye might be saved"—*ἵνα σωθῆτε*. On the other hand, the passage we are considering in the Revelation supplies an example of a fixed purpose carried into effect, *as the context proves.*

Dr. Wordsworth does not seem to have observed the clause which nullifies all his statements. It is *not* said that the angel laid hold of Satan *in order that he might* bind him.

It is said that the angel laid hold of Satan "*and bound him*," and that "for a thousand years." The accomplishment of the binding proves the certainty of that which is made dependent on the binding, viz., the non-deception of the nations during the appointed period.

May I be permitted here to say that Bishop Wordsworth should be more careful in his criticisms. When speaking of the two Witnesses (Rev. xi.) he argues that they could not be two individuals because it is said that "their body" (πτωμα αὐτων), not their bodies, "lay in the street," etc. If he had examined the Hebrew of Psalm lxxix. 2, which text he quotes, he would there find the same form of expression. "They have given the dead-body [אֶת־נִבְלַת singular] of Thy servants [plural] as meat," etc. It is a form of expression continually used in the Hebrew Scriptures, and adopted in the Greek. Thus in one chapter of Leviticus alone, Lev. xi., there are no less than nine verses in which we find the expression, "*their dead-body.*" So Deut. xiv. 8. "*Their dead-body* ye shall not touch." See also Isaiah v. 25. "*Their dead-body* was as the mire in the streets." In Ps. lxxix. 2, the singulars, "*corpse*," "*bird*," and "*beast*," are all collectives. In the New Testament we find abundant examples of the same form of expression. See Rom. iii. 13, ὁ λαρυγξ αὐτων, 14, ὡν το στομα; Matt. xviii. 16, δια στοματος δυω μαρτυρων, by the mouth of two witnesses; Acts iii. 21; Eph. iv. 29; Col. iii. 8; Rev. ix. 17, ix. 18, xi. 5; 1 John iii. 20; 2 Cor. iii. 15, ἐπι την καρδιαν αὐτων. Nothing can be more frequent than this usage of the singular both in the Old and New Testaments. That the expression το πτωμα αὐτων is to be understood in a plural sense is proved by the fact that the expression τα πτωματα αὐτων (plural) is in the very same verse used as synonymous with το πτωμα αὐτων. See Rev. xi. 9.

The two Witnesses were *officially* one—one in testimony. Individually, however, they were distinct. Hence πτωμα and πτωματα are both used of them.

§ IV.

IT will be seen from a preceding chapter* that the doctrine of "The First Resurrection" is not taught merely in an isolated passage of the Apocalypse, as is very commonly affirmed: it is frequently taught both in the Old and in the New Testament, and in no passage more clearly than in 1 Cor. xv. 23. There the Apostle reveals the order ($\tau\alpha\gamma\mu\alpha$) of the Resurrection of Christ and His saints. "Every one in his own order: Christ the first-fruits; afterward [$\overset{.}{\epsilon}\pi\epsilon\iota\tau\alpha$] they who are Christ's at His coming. Then [$\epsilon\hat{\iota}\tau\alpha$] cometh the end," when the last resurrection takes place.† Thus there are three periods of resurrection, and three only. First, Christ's: secondly, the resurrection of those who are Christ's at His coming, which is the *first* resurrection of the saints: thirdly, the final period, when all the saints who have lived during the Millennium will rise, as well as the rest of the dead of all generations. The

---

* See Chapter XIII., page 274.

† 'Eπειτα, *afterwards*, and ειτα, *then*, are particles of sequence, and are equivalent to *secondly*, *thirdly*. See previous remarks on this subject. The First Resurrection is a resurrection *out of* the dead, implied by the preposition ἐκ.

A A

First Resurrection is also taught wherever the expression ἡ ἐξαναστασις ἡ ἐκ νεκρων, or ἡ ἀναστασις ἐκ νεκρων is applied to Christ's people: as also whenever we find the expression, "saints of the high places." The evidence afforded by these, and such like passages, is irresistible; yet it weighs as nothing with Bishop Wordsworth. Being, no doubt, conscious that the doctrine of the millennial reign is so intimately associated with that of the First Resurrection that they stand or fall together, and being prepared to reject the inspiration of the Apocalypse rather than admit the truth of the millennial reign, Bishop Wordsworth finds himself under the unavoidable necessity of repudiating the doctrine of the First Resurrection as taught in the Apocalypse, and (it must be added) as taught in the rest of Scripture.

Let us consider his arguments. He first quotes Rev. xx. 4, 6. "And I saw thrones, and they sat upon them, and judgment was given unto them: and I saw the souls of them that were beheaded for the witness of Jesus, and for the word of God, and which had not worshipped the beast, neither his image, neither had received his mark upon their foreheads, or in their hands; and they lived and reigned with Christ a thousand years. But the rest of the dead lived not again until the thousand years were finished. This is the first resurrection. Blessed and holy is he that hath part in the first resurrection: on such the second death hath no power, but they shall be priests of God and of Christ, and shall reign with Him a thousand years."

"These words," says Bishop Wordsworth, "are "*not* spoken of the *bodies* of the saints, but of their "*souls. I saw the souls of them who had been be-* "*headed for the witness of Jesus.* This must be care- "fully borne in mind, because the error of the mil- "lenarians is mainly due to neglect of this distinc- "tion. They have imagined a *bodily* resurrection, "whereas St. John is speaking of a *spiritual* one." (Page 53.)

How could anything else than a bodily resurrection be indicated by the sight of the souls of beheaded persons living again? If beheading took from them their bodies and brought them into a state of death, their restoration to a state of life necessarily implies the restitution of their bodies. The beheaded must have been possessed of bodies once, otherwise they could not have been beheaded; and after they had been beheaded, their souls, if seen by the Apostle at all, must have been in a disembodied state in order that he might see them brought out of that state and "LIVE," that is, become re-united to their bodies—"*live*" being a word used only of those who are in the possession and exercise of the proper powers of their being: and those powers are not possessed by any whose bodies are in the grave. Disembodied souls exist, and are conscious (see Rev. vi.), but they are not said to "*live.*" All who are separated from their bodies are regarded as being in a state of death, not of life, till their bodies are restored to them. Accordingly, as soon as the Apostle had seen them "LIVE," that is restored to the possession of their bodily powers, we find the words, "THIS IS THE FIRST RESURRECTION"—resurrection being a word used of the resuscitation of the body, but *not used in Scripture of the quickening of the soul through the Spirit.**

Bishop Wordsworth, however, heeds none of these things. The words "that *had been* beheaded," seem to present no difficulty to him. He thinks that when the souls of those "that had been beheaded for the witness of Jesus" were seen to live and to reign, the lesson taught is, the entrance of the baptised into that new life pertaining to the mystical body of Christ, of which Baptism is the door. (Page 57.) Now, even if we were to admit that those who have been duly sprinkled with the mystic waters of baptism by priestly hand, were thereby brought into the glory and blessedness of the First Resurrection, yet it would seem difficult to explain how any one *who had been beheaded* could be either baptised, or quickened with spiritual life. How could a beheaded person be baptised? Obviously, it is necessary for Bishop Wordsworth either to cancel the words "had been beheaded," or else

* It is most important to remember this. The rejection of this truth involves grave doctrinal error.

to abandon his theory of Baptism bringing into the blessedness of the First Resurrection. When too we think of the course which baptised Christendom has trodden, there seems something very awful in its being said that baptism entitles any to ascribe to themselves the blessedness which these solemn words declare. "Blessed and holy is he that hath part in the first resurrection: on such the second death hath no power, but they shall be priests of God and of Christ, and shall reign with him a thousand years."

Bishop Wordsworth is not very likely to renounce his doctrine respecting Baptism, for it is the very pillar of his system. But if he should see the absolute impossibility of saying that those who had been *beheaded* could after such beheading be either baptised or spiritually quickened, and if he should further acknowledge that the souls of the *beheaded* must be *disembodied* souls, he would, in that case, be obliged to abandon the thought that "*live*," when applied to such disembodied souls, could mean that they were quickened with *spiritual* life: for spiritual life cannot be communicated to souls after death; nor, secondly, could any have been "beheaded for the witness of Jesus, and for the word of God" who had not been spiritually quickened whilst yet in the body. Persons dead in sin could not bear witness for Jesus.

Millenarian notions are, according to Bishop Wordsworth, ascribable to low and inadequate notions of our baptismal privileges and obligations, and of the sacred duties and inestimable blessings of Church membership and Church unity; and wherever unworthy notions are entertained on these momentous points, *there* the doctrine of the Millennium may be expected to prevail.

"Our spiritual adoption into the mystical body
" of Christ [I continue to quote the words of Bishop
" Wordsworth] is only the *beginning* of our Christian
" life; it is the new *birth*, that is, it is the *entrance*
" into the new life. Baptism is the door by which
" we enter Christ's Church. But the door is not the
" house. [Yet baptism, according to Bishop Wordsworth, brings us through the door *into the house*, and that house, a house of present glory in the earth.]
" There must not merely be new *birth*, but a new
" life. [But he who has the new birth must have new life; and it is the possession, not the development of this life which, according to Bishop Wordsworth, brings into the First Resurrection; and the possession of that life was, he says, attained by
" baptism.] There must be not only the *mark* of
" Christ imprinted on the forehead, but there must
" be the *spirit* of Christ moving in the heart, and
" bringing forth the *work* of Christ in the hand.
" This is the First Resurrection. We rose with
" Christ, [in baptism] to live with Christ; and Christ
" assists us in this work by manifold gifts and
" graces. In the Holy Communion of the body and
" blood of Christ, the Christian soul receives spiritual

"strength from Him, and is knit more closely to
"Him. There we dwell in Christ, and Christ with
"us. We are one with Christ, and Christ with
"us. Thus the soul which was born again in
"Christ, [by baptism] *lives* with Christ, [by means of
"the Eucharist.] It is dead to sin, and is ready to
"suffer for Christ, and knows no other object of
"worship than Christ. [Has this been so in sacer-
"dotally baptised Christendom?] *I saw the souls of*
"*them that were beheaded for the witness of Jesus,*
"*and which had not worshipped the beast, neither his*
"*image, neither had received his mark in their fore-*
"*heads, or in their hands*—that is, who had not
"broken their oath of allegiance to *Christ*, either in
"word or deed: *and they lived and reigned with*
"*Christ a thousand years.* This is the First Resur-
"rection."

Such are the statements of Bishop Wordsworth. First, we are told that men are brought into the blessedness of the First Resurrection by baptism. Now, we are told something very different. We are taught that those brought into this blessedness were those who had been beheaded for the witness of Jesus, had testified against the beast, and his image, and "had not broken their oath of allegiance to Christ either in word or in deed." Are these things pre-requisites to baptism? If so, how was it that the Pentecostal Church was baptised, for they certainly did not testify against the beast, or his image, nor had they had any time to prove the sincerity of their allegiance to Christ, seeing they were baptised the moment they believed. If baptism be the door, then these things cannot be the door; or if these things be the door, then baptism cannot be the door. Can Bishop Wordsworth explain which he means, or has he involved himself in a labyrinth from which there is no extrication?

To those who understand the Scripture I need not say, that Bishop Wordsworth has no apprehension whatever of what the blessedness, holiness and glory of the First Resurrection really is. If he had, he would never have dreamed of saying that we are by baptism brought into it. He is utterly ignorant of what the Scripture declares respecting that now approaching period when "the kingdom, and dominion, and the greatness of the kingdom UNDER THE WHOLE HEAVEN shall be given to the people of the saints of the Most High," (Dan. vii.,) and "the sovereignty of the world become the sovereignty of our Lord and of His Christ." Rev. xi. The national conversion of Israel, and their establishment as God's saintly nation in the earth; the binding of Satan, so that he shall cease for the thousand years to deceive the nations; the release of creation from its present groan; the cessation of strife among the nations, so that they "beat their swords into ploughshares, and learn war no more;" the extinction of savage passions in the animals, so that the lion shall eat straw like the ox, and the wolf and the lamb lie down together; the resurrection of the saints into glory above the heavens, whence they will reign with Christ over all things; the continuance of this Adamic earth for a thousand years; the occurrence at the close of that period of a brief apostasy of the nations; the termination of the history of the Adamic earth, and the creation of a New Heavens and a New Earth in which righteousness, and only righteousness will dwell— all these characteristics and results of the coming reign of Christ are, by Bishop Wordsworth, unrecognised and unknown.

Bishop Wordsworth's description of the reign of the saints is as follows:—"*I saw thrones, and they* " *sate upon them, and Judgment was given them.* Our " Blessed Lord expressly says, that the Judgment of " Satan was already begun at His own Incarnation. " *Now is the judgment of this world. Now is the* " *Prince of this world judged.* And now, even now, " the Saints of Christ judge the world; yea, accord- " ing to St. Paul's words, *they judge Angels*, the " Angels of Satan. The Saints of God prove, by " *their* faith and holiness and steadfastness, that the " fall of Satan and of his Angels was due to their " own sin : they show, by their virtues, that God is " good, and that His grace is sufficient for all those " who pray to Him, trust in Him, and obey Him; " and that *it is made perfect in their weakness, and* " *that His commandments are not grievous ; and hav-*

"*ing been tortured, tempted, afflicted, tormented,* and
" *having resisted even unto blood, and having come*
" *forth more than conquerors,* they judge the world.
" They condemn it of blind infatuation, and of base
" ingratitude to God. The life and death of the
" Saints is the judgment of the world.

"Again, in another sense, the Church of Christ
" now judges the world. She has received from
" Christ the power of *the keys;* the power of *bind-*
" *ing and loosing;* and whatever she does on earth,
" orderly and rightly, in the ministry of remitting
" or retaining sins, is ratified by Christ *in heaven.*
" Thus, *even now,* the Saints of God sit upon thrones,
" and *to them judgment is given.*

"Yet more; in another manner the Saints of God
" are even now seated upon Thrones, and judge the
" world.

"In the precepts of the Law, in the revelations of
" Prophets, in the melody of Psalms, in the instruc-
" tion of Proverbs in the Old Testament, the Twenty-
" four Books of which were believed to be repre-
" sented by the Twenty-four Elders sitting enthroned
" in heaven; and in the four Gospels typified by the
" four living Cherubim on which the Throne of God
" is set; and in the *Royal Law* of the Letters of
" the Apostles, whom God has *made Princes in all*
" *lands;* which books, be it remembered, have been
" placed on Thrones in the great Council-halls of
" Christendom, and have been delivered as a Law to
" anointed Kings at their solemn enthronisation; yes,

"taken from that very altar,* and placed in the
"hands of the most august Monarchs of the world,
"in this national Temple, at their Coronation; and
"whose sanctity is proclaimed by solemn adjurations
"in Courts of Justice; and which are delivered to
"Bishops and Priests at their Ordination, as the
"Royal Code of their Teaching, and the Divine
"Charter of their Ministry; and which sound forth
"daily from Pulpits and the steps of Altars—as it
"were, from Christian Thrones and Tribunals—in
"every part of the world: thus, I say, they whom
"God has employed to declare His Will to men,
"are now, even now, seen by the eye of Faith *sitting
"*upon Thrones; and to them Judgment is given.*

"In this manner we see that the souls of the
"Saints, by virtue of their spiritual incorporation
"and indwelling in Christ, have risen from the dead
"with Christ; that in Christ they live; that they
"ascend with Him, and sit with Him in heavenly
"places; that they are Priests of God and Christ;
"that they reign together with Him; and that with
"Him they judge the world. Therefore—*Blessed and
"holy is he that hath part in the first resurrection.*"

Thus, that which was the Church's curse, shame, and apostasy, in that it refused to follow the steps

* These sermons appear to have been preached in Westminster Abbey as well as before the University of Cambridge. It should be observed that wherever there is a deliberate substitution of the word "*altar*" for "table," there is a virtual abjuration of Protestantism.

of the Lord Jesus, and resolved, laying aside the garb of Nazareth, "to reign as kings" whilst Satan continued to be the god of this world—that step of apostasy taken immediately after the Apostles died, ratified in the time of Constantine and not since repented of, Bishop Wordsworth glories in. He calls it reigning with Christ. He should rather have said, reigning with Satan. Even before the Apostles died, the disposition to exalt themselves in the world by means of God's Truth, and to reign as kings before the time, was manifested; but the Apostles checked it. "Already ye are full," said the Apostle Paul to the Church at Corinth, "already ye are rich, ye have reigned as kings without us: and I would to God ye did reign, that we also might reign with you..... We are fools for Christ's sake, but ye are wise in Christ; we are weak, but ye are strong; ye are honourable, but we are despised. ..... we are made as the filth of the world, and are the offscouring of all things unto this day." Such was the place which the Apostles held, and while they lived, the Churches were not permitted quite to abandon it.

There was a period in the history of the Lord Jesus when the multitudes wished "to come and take him by force and make him a king." They had discovered that He was possessed of mighty powers, and they deemed those powers available for the promotion of their worldly interests. They wished Him to sanction their principles, and to become their leader in the vicious paths that they were seeking to tread. In a word, they desired Him to take the headship of human things *as they at present are.* If He had consented to do this, they would gladly have owned Him as their Lord and Master. But He consented not. He came to bear witness for God and for His Truth, and ·to declare that "the foundations of all things were out of course"; that, consequently, *essential change* was needed, not modification. When, therefore, they found that He would not yield to their seductions, or sanction their principles, and that there was, as it were, "a hiding of His face from them,"* they despised, hated, and at last crucified Him. One aspect of the Cross is, that it is the abiding memorial of the distance which separates unregenerate man from God. Man reared the Cross, and nailed thereunto the only One in the earth in whom God delighted. They crucified One who was indeed man, holy, perfect, and blessed, but One who was also God their Creator. So far from having been brought nearer to God either by the Incarnation, or by the Death or by the Resurrection of Jesus, the world has gone into further distance from Him, for it has seen, considered, and spurned Him and His Truth. "The whole world," said the Apostle, "lieth in the Wicked one." This was true before the Incarnation, and it is true still. The

* See these words considered in "Thoughts on Scriptural Subjects," page 192, as advertised at end.

Church was chosen *out of* the world in order that it might bear witness to the fact of the world's condition, and also that it might testify to *grace*. All true believers (not those who may, like Simon Magus, have been baptised) having a legal oneness with Him who obeyed, died, and rose *in their stead*, and having also a living union with Him through the Spirit, are *representatively* in Heaven, and in that sense are exalted: but they are not *personally* in Heaven. Personally they are on earth, appointed for the present, not to reign, but to suffer: to be persecuted even as their Lord was persecuted. To know "tribulation" and "endurance" is the present portion of those who are "in the kingdom in Jesus;" yea, to seek as their highest honour not to reign as kings, but to lose their lives in this world, and to suffer. For a little while, the Churches walked in this path; but when the Apostles died, they corrupted themselves. They sought influence and aggrandisement in the world by means of Christ's Truth. They adulterated the doctrines, and lowered the principles of Christ into adaptation to the world's thoughts. Adulterated Christianity soon won the world's favour; for it was found serviceable to men's *present* interests. Accordingly, to debased and fallen Christianity, kingship was tendered, and thus, to use the expression of a modern writer, "*Christianity ascended the throne of the Cæsars.*" It enthroned itself and apostatized. Nor has the sin been repented of. On the contrary, it is rejoiced in and extolled; and although not frequently magnified in words so jubilant as those I have just quoted from Bishop Wordsworth, yet the general sentiment of Christendom has been, and is, that the Church is at present enthroned, or designed to be enthroned, on Zion and reigning with Christ. Orientals, Papists, Protestants, Conformists, and Nonconformists, have alike welcomed the delusion and fallen into the deadly snare. Nor will Christendom as a whole ever repent of their early sin. A remnant, as in Israel of old, will be gathered out; but Christendom as a whole will not retrace its steps. A large portion of it—all that is included in the name Ἡ Οἰκουμενη, will avowedly reject Jehovah and Christ, and gather around Antichrist. The rest of Christendom, though it will not in the same way governmentally apostatize, will nevertheless teem with infidelity and false profession. "Tares," "bad fishes," "foolish virgins," "wicked servants," "goats," "fruitless branches" in the vine, will abound. The nations have never ceased to be deceived by Satan. They have never truly been brought to Christ. His true servants have ever been "a little flock"—"lambs in the midst of wolves." The path of "the just" will surely not widen: it will become more and more narrow, as the closing hour comes on.

§ V.

WILL "the morning without clouds" ever arise upon this groaning earth? *That* is a solemn question. We have seen the way in which the Teachers of Christendom are wont to answer it. They reply, "It will never arise in any other way than that in which it has already risen. The true light has already come. Has it not shone? And is it not shining still"? Yes; it has shone. "The Light shineth in darkness, and the darkness comprehended it not." The darkness of *Heathenism* has not departed. The darkness of *Judaism* has not departed. The darkness that has come in over *Christendom* has not departed. On the contrary, it is increasing every day; as the present condition of our own once favoured country abundantly testifies. Before "the morning without clouds" can come, the earth must be brought under the present controlling power of One that is "*just, ruling in the fear of God*"—One who shall not only be strong enough to grasp "the sons of Belial," but who, after having grasped and gathered them, shall cause them to be "utterly burned with fire in the same place." "The sons of Belial shall be all of them as thorns thrust away, because they cannot be taken with hands; but the man that shall touch them must be fenced with iron and the staff of a spear; and they shall be utterly burned with fire in the same place." (2 Sam. xxiii. 6, 7.) The sons of Belial are not yet thrust away and burned. Their strength increases, and will increase greatly. He whose coming is to be as the light of the morning when the sun riseth has not yet returned in the greatness of His strength to grasp them; and until He shall thus return, "the morning without clouds" cannot be. "Watchman, what *of the night?* Watchman, what *of the night?*" is still the enquiry amongst God's people. "The morning COMETH" is part of the reply; but, it is added, "*also the night.*" We have long been in the night, but its darkest hour—the hour that immediately precedes the dawn, is yet to come. There have already been "many antichrists"; but "THE ANTICHRIST" has not yet been revealed. It is well for us to remember that if, of old, Israel took the lead in rejecting Christ, Christendom is now foremost in preparing the way of Antichrist. The ten kings that are to "have one mind," and "agree to give their kingdom and power" to him, are *Gentile* Kings. They will be kings whose kingdoms are to be formed out of H OIKOYMENH, the Roman World; and the nations of the Roman World once were, and many of them still are, the very centre and soul of Christendom.

The Western nations of the Roman World are now engaged not only in "*changing*" (I do not say, *re-modelling*) their own institutions, but they are intent also on resuscitating those Eastern countries whence Western civilization was derived. They are re-entering the regions in which the Patriarchs, and the Prophets, and CHRIST, and the Apostles testified. They are re-entering that Land which God has claimed as peculiarly His own and named it Immanuel's Land. They are re-entering the lands which God has prophesied of by name—lands which He has already, because of past iniquities, smitten, but which are again to become the platform of the last developments of the counsels of men — lands whose course, and whose doom, God has pourtrayed with faithful accuracy in His holy Word. Those lands Western Europe is now re-entering to mould them anew, and is fashioning them according to its own principles. And what are those principles? God has given them their name. He calls them principles of Lawlessness (ANOMIA), that is to say, a lawless licentiousness of thought and action, that boasts in having set itself free from all restrictions *that come from Him.* Men will say both of Jehovah and of Christ, "Let us break their bonds asunder, and cast away their cords from us."

Sometimes in the history of the world incidents occur, trivial in themselves and transitory, but not without significance to those who watch the signs of the times in the light of the Word of God. A few years ago the reign of Ecclesiasticism and Mediævalism in Rome suddenly collapsed. The power which Ecclesiasticism had so long wielded over that city was transferred into other hands. Sacerdotalism retired, and Secularism entered— entered on an inheritance whose condition seemed fitly to indicate the character of the rule of its former lords. Storm, tempest, and devastating floods had visited Rome. The Tiber had overflowed its banks, and was rushing through the streets of the city. Famished and homeless multitudes were wandering in search of food and shelter; but there was no kindly care extended, no voice to direct, no hand to succour. On a dark and stormy night, summoned by their cry of anguish, the new Monarch of Rome entered. As soon as the morning dawned he went forth to view the scene, and took his stand in front of the Capitol, there to inaugurate his reign.

Behind him therefore, was the symbol of the departed greatness of *Pagan* Rome—the memorial of a period of darkness, misery and evil, the real character of which eternity only will unfold. Before him stood the yet unruined trophies of Rome's Ecclesiastical greatness; but over them all darkness brooded, and from them the wail of distress came. An instructed eye, whether looking back upon the past, or contemplating the present, could see nothing but misery there—misery for time—

misery for eternity. Such was the scene in the midst of which Rome's new Deliverer appeared. Rome's ancient history, and Rome's mediæval history had closed. Victor Immanuel could truly say to the stricken multitude that stood before him,. "The past has brought to you no prosperity. It has brought you no deliverance from desolation, and misery, and woe. Ye need a deliverer." What if he had taken the Bible in his hand and said, "Let Him who wrote this book be your Deliverer. Seek ye unto Him. Pour out your heart before Him, ye peoples. God is a refuge for us." But he spake not, he thought not, of this. He stood there as the impersonation of modern principles, that he might present himself acting in the power of those new principles as their Deliverer. His words virtually were, "Follow me as your leader and guide, and I will show you what human energy can effect when its bonds are rent, and its proper liberty of thought and action restored." Self-government is the idol of the day. It is one of the forms under which man worships himself. It is being tried in the West; it is to be introduced into the East as the panacea of the evils under which Syria, Asia Minor, Greece, and Egypt, have so long groaned. But self-government is not the government of God. Men even if perfect, as the angels are, would still need to be governed. But what must self-government entail when men are not only not perfect, but instinct with various forms of evil, more numerous than the hairs of the head, and surrounded also by spirits of darkness whose multitude no man can number?

England has been the country in which self-government has been longest tried. And to what is it leading England? There seemed to be disguised, but bitter irony, in the words recently addressed by the Prime Minister to the House of Commons when he reminded them that having retired from the positions which they once held —first, as an assembly of Churchmen, then of Protestants, then of Christians, they need not hesitate to carry out their principle to its legitimate conclusion, and profess themselves to be, *religiously*, nothing. "The narrow ledge of Theism" was all that was left, and was it worth while to fight for the maintenance of that? Expediency is the great rule of natural and social life now; therefore, if expediency requires that atheists should be admitted as legislators, it must be so. England consents to accept as one of her future axioms that men who deny the existence of God, and spurn the Bible, are as well, it may be better qualified to govern, than those who fear God, and own the authority of His Word. Daniel and Paul did not think so. Else they would not have said what they did say to Nebuchadnezzar and to Belshazzar, to Festus and to Agrippa. Yet both Daniel and Paul spoke as the divinely-inspired messengers of God. But who or what is God, and what the value of His Word, when compared with the majesty, and dignity, and scope of modern thought and "the higher culture?" So saith the fool. Society listens, and obeys.

---

§ VI.

IF it be asked what will the end of these things be, the Scripture supplies an unambiguous answer. The present efforts of Western Europe to resuscitate the East, and then to indoctrinate it with modern principles, will *ultimately* succeed. The unerring (though despised) voice of God has declared that a system of governmental lawlessness (ἀνομια) is to be *established* in the land of Shinar.* That such a system has been long in process of construction in the West is too obvious to need proof. The question, whether the force of the rules and principles that are to guide society, is to be founded on the authority of God, or merely on the authority of man, has been long considered, and been determined in favour of man. The ignorement of laws, natural and revealed, hitherto recognised as proceeding from God, and the ignorement in many cases of the fact of His existence, must necessarily introduce a radical change into the order of society. The principles of this new and

* See Zech. v. 5, as considered in "Prospects of the Ten Kingdoms."

godless system may be said to be accepted now, though its rules are not yet matured and codified. The time is not yet fully come for their embodiment and promulgation, and when it has come, the promulgation will be from "the Land of Shinar," where the recognised Areopagus of the Ten Kingdoms of the Roman World will be. Assyria and the Valley of the Euphrates will soon become the great sphere of human action. Those names will soon be as familiar to the ears of men as Greece and Egypt, Athens, and Cairo, now are.

At present, however, the way is being *prepared* merely. "The Transgressors" (see Dan. viii. 23) are not yet "come to the full," nor will they, until apostatising Europe has first practised upon Israel, and drawn them, as a nation, into avowed association with its godlessness. Israel is not quite prepared for such association now. Although *unconverted*, when first they return to their own Land, they will nevertheless stand (for a time at least) in favourable contrast with the empty pretentious Scepticism, and the debasing Idolatries that are now spreading over Europe. They will not consent to disown Jehovah as their God; or to charge untruthfulness on Moses and the Prophets. They have not yet forgotten that David has said, "The fool hath said in his heart, There is no God"; and the Psalm that contains those words, they will, no doubt, chant in their Temple with a keen sense of its being before their eyes fulfilled. Nevertheless, having rejected the One Shepherd, they will have with them no adequate power to cope with the Destroyer. The subtilty and blandishments of European Antichristianism will surround them. On them too the power of the coming delusion will in all its intensity fall, and they also shall believe the last great lie of Satan, and be even foremost in sustaining it. When the apostate Jew shall join with the apostate Gentile (whether such Gentile be nominally Christian or Heathen) in avowedly repudiating the authority of an unseen Supreme Ruler greater than themselves, the Transgressors will have come to the full; Lawlessness will have been enshrined in the Land of Shinar; and THE ANTICHRIST will have been revealed.

How marvellous the blindness that has for ages rested on true Christians, so as utterly to prevent their recognising the future of Assyria as delineated in the Word of God. If they think of Assyria at all, they think of it as a country that once was, but which has vanished now, and sunk into the grave of ages never to rise again. Yet Assyria is to be revived—revived first by man in order to defy God in a manner in which no nation or country has ever yet defied Him: afterward it is to be revived again under the hand of God, in order that it may serve and glorify Him, and become a blessing in the earth. Although it will be one of the chief agents of Satan in the coming night of evil: it will also be one of the chief instruments of God for good when "the morning without clouds" shall have come.

Read the nineteenth chapter of Isaiah. The bruising and healing of EGYPT is there first spoken of. Egypt has not yet been healed. When once it has been healed, it will be healed effectually and for ever. A terrible blow more crushing than any it has yet received, impends over Egypt. Yet, even Egypt is to be finally forgiven. Its healing is mentioned; and then the chapter proceeds to speak of *Assyria*. "In that day shall there be a highway out of Egypt to Assyria, and the Assyrian shall come into Egypt, and the Egyptian into Assyria, and the Egyptians shall serve him with the Assyrians. In that day shall Israel be the third with Egypt and with Assyria, even a blessing in the midst of the earth [הארץ]: whom the Lord of hosts shall bless, saying, Blessed be Egypt my people, and Assyria the work of my hands, and Israel mine inheritance." This will be when "the morning without clouds" shall have come.

But Assyria will have a dark and awful history first. Assyria and her last great Monarch (soon to appear) will be at the head of the Apostasy that ends the history of this evil age. Asshur [אשור] is the great denominative title of Antichrist in the Old Testament Scriptures. Sometimes he is called "the King of Babylon"; sometimes "Asshur." These names are used interchangeably, even in the same chapter. (See Isaiah xiv.) As the Head of Assyria he leads the last great Apostasy against God; seduces, and then betrays the Jews; tramples them down as the mire of the streets; places his Idol on the pinnacle of their Temple; destroys in Jerusalem the two Witnesses; sacks Jerusalem, and carries half the people into captivity; again assaults Jerusalem with the view of blotting it out utterly, that the name of Israel might be no more had in remembrance, when, suddenly, the great Head of Israel will interfere, forgive and rescue Jerusalem, and the kingdom of God will come. "Let the Gentiles [הגוים] be awakened, and come up to the valley of Jehoshaphat: for there will I sit to judge all the Gentiles round about. Put ye in the sickle, for the harvest is ripe: come, get you down; for the press is full, the fats overflow; for their wickedness is great. Multitudes, multitudes in the valley of decision: for the day of the Lord is near in the valley of decision. The sun and the moon shall be darkened, and the stars shall withdraw their shining. The Lord also shall roar out of Zion, and utter His voice from Jerusalem; and the heavens and the earth shall shake; but the Lord will be the hope of His people, and the strength of the children of Israel. So shall ye know that I am the Lord your God dwelling in Zion, My holy mountain: then shall Jerusalem be holy, and there shall no strangers pass through her any more."

One of the most remarkable descriptions of the glory and fall of the great Head of Assyria is found in the *fourteenth* chapter of Isaiah—a chapter the more to be noted, because the most careful expositors of Isaiah, such as Vitringa, Bishop Lowth, and Bishop Horsley, admit that that chapter has not yet been fulfilled. They admit that it gives us a prophetic sketch of the course and fall of the last great Head of the Gentiles. In admitting this they virtually admit the revival of Assyria and Babylon, as well as the futurity of all other chapters (such as Isaiah x.) which speak of the same personage, and the same period—the period being marked by its being the time when God rescues, and forgives Israel, and restores them to His favour for evermore. Israel is not yet forgiven. The indignation of the Lord against them hath not yet passed. They are not yet "blessed in basket and in store, blessed in their going out and in their coming in." All nations do not yet "call them blessed." "It shall come to pass in the day that the Lord shall give thee [Israel] rest from thy sorrow, and from thy fear, and from thy hard bondage wherein thou wast made to serve, that [*then*, not before] thou take up this proverb against the King of Babylon, and say, How hath the oppressor ceased! the golden city ceased!" (Isaiah xiv. 3.) The rise, therefore, and glory, and fall of this great Oppressor must be accomplished before Israel is forgiven. He it is who shall head "the sons of Belial who are to be gathered and utterly burned with fire in the same place." "The Lord of hosts hath sworn, saying, Surely as I have thought, so shall it come to pass; and as I have purposed, so shall it stand: that I will break THE ASSYRIAN in my land, and upon my mountains tread him under foot: then shall his yoke depart from off them [Israel], and his burden depart from off their [Israel's] shoulders. *This is the purpose that is purposed upon the whole earth: and this is the hand that is stretched out over all nations.* For the Lord of hosts hath purposed, and who shall disannul it? and His hand is stretched out, and who shall turn it back?"

Will any one affirm that any event has ever occurred in the Land of Israel, or anywhere else throughout the wide earth, that has resulted in the forgiveness of Israel, and in the accomplishment of the purpose for which from the beginning, the hand of Jehovah has been "*stretched out over all the nations.*"

It will be an awful hour in the world's history when the lips of God shall really say, "*Ho, Assyrian,*" and so give him a charge against Israel "to take the spoil, and to take the prey, and to tread them down like the mire of the streets." (See Isaiah x.) The kings of the whole Roman World (της οἰκουμενης ὁλης) and their armies, will by Him be summoned to Armageddon in the vicinity of Carmel. To this mighty power a vast addition will suddenly be made by the accession of the Arabians and others of the kindred of Israel. The enumeration of these added nations is thus given in the eighty-third Psalm. "The tabernacles of Edom, and the Ishmaelites; of Moab, and the Hagarenes; Gebal, and Ammon, and Amalek; the Philistines with the inhabitants of Tyre; Assur also is joined with them; they have holpen the children of Lot." (Ps. lxxxiii. 6—8.) The object of this mighty confederation—more mighty than any that the world has ever yet seen, will be the utter destruction of Israel and of Jerusalem. "Come, and let us cut them off from being a nation; that the name of Israel may be no more in remembrance." See verse 4. With this object THE ASSYRIAN will advance upon Jerusalem. His progress is recorded thus. "He is come to Aiath, he is passed to Migron; at Michmash he hath laid up his carriages: they are gone over the passage: they have taken up their lodging at Geba; Ramah is afraid; Gibeah of Saul is fled. Lift up thy voice, O daughter of Gallim: cause it to be heard unto Laish, O poor Anathoth. Madmenah is removed; the inhabitants of Gebim gather themselves to flee. As yet shall he remain at Nob that day: he shall shake his hand against the mount of the daughter of Zion, the hill of Jerusalem. Behold, the Lord, the Lord of hosts, shall lop the bough with terror: and the high ones of stature shall be hewn down, and the haughty shall be humbled. And he shall cut down the thickets of the forest with iron, and Lebanon shall fall by a mighty one." (Isaiah x. 28—34.) The proud power of the Gentiles shall fall, and shall never rise again. Are we to be told that this has already been? If so, the Image of which we read in Daniel would have been smitten, ground to powder, and scattered to the four winds of heaven.

And what is more, the chapter which follows, and which is really another section of the tenth, would have been fulfilled likewise. When the Assyrian falls, "the times of the Gentiles" will end, and the time will have come (to use the words of Daniel) for "the saints to possess the kingdom." Not till then shall "the rod from the stock of Jesse COME FORTH," and "the Branch from his roots FLOURISH," that is, He who is now withdrawn from the earth and hidden in the heavens shall COME FORTH, not again to be as "a root out of a dry ground," but "*shall come forth*" to flourish, and the result shall be righteous judgment in the earth, so that "the poor" and "the meek" (who now suffer) shall be vindicated and upheld. The earth also, which has hitherto been the subject of His long-suffering mercy, shall be SMITTEN, and the Leader of its rebellion destroyed. "He shall smite the earth with the rod of His mouth, and with the breath of His lips shall He slay The Wicked One." Even if the Apostle had not quoted and explained these words in

2 Thess. ii. 8, no one who had regard to facts could doubt the futurity of the event spoken of.

Wickedness, falsehood, and misery, shall fall: righteousness, truth, and peace, enter. Such is the subject of this chapter. The wolf shall prey upon the lamb no longer: they shall dwell in peace together. The lion, as in Paradise, shall feed on the fruits of the field. The Land of Israel, for ages the resting-place of lamentation, darkness, mourning, and woe, shall be "filled with the knowledge of the Lord, as the waters cover the sea." None shall hurt nor destroy in all God's holy mountain. Ephraim and Judah shall no longer vex each other, but shall be gathered back to the Land from which they have been driven, and dwell in peace together. The tongue or bay of the Egyptian sea which has ever severed, and still severs Egypt from the Land of Israel, shall sever it no longer; for it is to be dried up. Egypt is to be brought into close and blessed association with Immanuel's Land. So also is Assyria. "The river," that is the Euphrates,* is to be smitten (not *in*, but) "into [לְ not בְּ] seven streams, and men shall pass over it dryshod" [ἐν ὑποδήμασι, Sept., with their sandals on]. What nations hitherto have been more divided—

---

* The reasons for saying that "the River," which is to be "smitten *into* seven streams," means the Euphrates are elaborately given by Vitringa *in locum*. He observes "UBI נָהָר ἁπλῶς *occurrit passim usurpatur de Euphrate.*" See also Alexander *in locum*.

what more set in hostile and jealous rivalry against each other than *Assyria, Israel,* and *Egypt?* But when the time spoken of in the eleventh of Isaiah comes they shall be divided no longer. Linked in hallowed brotherhood, they shall be a blessing in the earth. " In that day shall *Israel* be the third with *Egypt* and with *Assyria*, even a blessing in the midst of the land: whom the Lord of hosts shall bless, saying, Blessed be *Egypt* my people, and *Assyria* the work of my hands, and *Israel* mine inheritance." (Isaiah xix. 24.) Shall we say that these things are accomplished? We *may say* so, if we please. Christendom *has* said so. It might just as truly have said that we are now in Paradise, and that Adam had never fallen. When the testimonies of God and the plainest evidence of facts are rejected, it can be of little use to argue. A cloud has settled in which nothing but the intervention of God's almighty power can remove.

If we ignore all that the Scripture has revealed respecting the specific characteristics of the present night of evil—if we ignore also all the agencies that are to introduce the morning of deliverance, and all the characteristics of the succeeding Day, what remains to us of Truth? Bishop Wordsworth's system affords no place for the history of Assyria, or of Egypt, or of Israel, either during the coming period of their evil greatness, or during the subsequent period of their blessing. The facts of Scripture are buried by him in the darkness of a wild Origenism. His principles would lead us back to the days of Constantine, and would make us what Eusebius was, a worshipper of Constantine and Constantine's principles. Yet what seed-principle of Popery is there that was not germinant, and more than germinant, in Constantine's time? You may divest Romanism of its political and practical configurations, yet doctrinal Romanism remains. Evidently, in Bishop Wordsworth's judgment, Satan was more manifestly bound during the age of Constantine than in any other part of the past thousand years of the saints' *supposed* reign. Armageddon, which Scripture places *before* the thousand years of blessing, Bishop Wordsworth places *after*. No unimportant alteration this. He thinks the gathering to Armageddon to be nigh. Satan will be loosed, and the Holy City, and the camp of the saints assailed, and that then all will end. What he thinks the Holy City and the camp of the saints to be, does not clearly appear. That his thoughts on that, as well as all kindred subjects, are utterly opposed to the Scripture, is, alas, too evident.

CHAPTER XIV.

REMARKS ON INDEFECTIBILITY.

INDEFECTIBILITY, that is, perpetuity of continuance in a condition of privilege or blessedness possessed, can only attach to creatures by the appointment of God. As He only can give, so He only can preserve blessedness that has been given. In Heaven, the condition of the elect angels is indefectible, because God has appointed that so it should be. In earth, the condition of man in Paradise *might have been* made indefectible; but it pleased God to appoint otherwise. No corporate body that has hitherto been constituted in the earth, has been called into an indefectible condition of blessing. Yet there is a body to which *the promise* of an indefectible standing *in the earth**

* I emphasise the words "*in the earth*," because I am not here speaking of the *invisible* Church, consisting of all the elect of all Dispensations from Abel down to the last individual that shall be gathered into the fold of faith during the Millennium. The standing *in Christ* of each *individual* believer is indefectible from the moment he believes; and the standing of the whole Church *corporately* will be indefectible, when they shall be gathered together as one glorified body in the new Heavens and new earth, in "the

C C 2

is given. It is promised, *not* to Gentile Christendom, for Christendom, like Israel before, has failed: it has not continued in God's goodness, and there-

---

Dispensation of the fulness of times." Of this I am not now speaking. What I have said is, that no *visible body* as yet constituted (whether Israel, or those gathered at Pentecost) had bestowed on them the privilege of indefectible standing *in the earth*—that is, no pledge was given them that they should corporately retain the privileged and honoured position in which they were at first set. Israel, who were first called into such a position, were cut off, and a similar *abscission* (ἀποτομια) was threatened to Christendom if it continued not in God's goodness. It has *not* continued in God's goodness. Scripture predicted this; and present facts attest it.

The words, "I am with you alway, even unto the end of the age" (αἰωνος), were addressed to the Apostles—not to the Church, or Churches gathered by them. The Apostles were the *Legislators* of the Church. In that office they have no successors. Their testimony was *inspired* testimony. It was committed *to writing*, that it might be an abiding testimony, and by it Christ works every day, and will continue so to work, even "till the end of the age." No new testimony will be sanctioned by Him as authoritative.

The false interpretation of this text has deceived multitudes unto destruction. If there were a body now existing in the earth to whom God had granted an authoritative and indefectible standing, it would be an act of rebellion against Him if we did not join it: but it would be no less an act of rebellion to join a body that falsely assumed such authority. To own a false Church, is as great a sin, as to own a false Christ.

The Church of Rome is a body that has so rioted in iniquity, as to make even unregenerate men shudder. To

fore shall be "cut off." The body to which
it is promised is *Israel*, when the day of their
blindness having passed, they shall be brought
as a nation, under the New Covenant of grace.
"The Redeemer shall come to Zion, and unto
them that turn from transgression in Jacob,
saith the Lord. As for Me, this is My covenant with them, saith the Lord; My Spirit that
is upon thee, and my words which I have put
in thy mouth, *shall not depart out of thy mouth,
nor out of the mouth of thy seed, nor out of
the mouth of thy seed's seed, saith the Lord, from
henceforth and for ever.* Arise, shine; for thy light
is come, and the glory of the Lord is risen upon
thee. . . . *Thy sun shall no more go down; neither
shall thy moon withdraw itself: for the Lord shall
be thine everlasting light*, and the days of thy

---

ascribe to such a body the attributes of an unfallen Church,
is an act either of wicked perverseness seeking to deceive,
or else an act of madness. Was it said to the Church in
Ephesus, simply because it had left its first love, "I will remove thy candlestick out of its place, except thou repent,"
and has no similar sentence been pronounced over a Church
which once shone as a lamp of God's sanctuary in Rome,
but now stands there as a synagogue of Satan?

They who desire to partake of the blessing promised to
the Apostles, must seek to abide in the Truth which they
ministered. They who depart from that Truth, and teach
other things, even though they be true believers, must expect chastisement—not blessing.

mourning shall be ended. Thy people also shall be *all righteous:* they shall inherit the land for ever, the branch of my planting, the work of my hands, that I may be glorified." (Is. lix. and lx.) These words are conclusive as to the indefectibility of the standing of converted Israel. They "shall be ALL righteous: they shall inherit the land for ever." The words which God shall put into their mouth shall not depart out of their mouth for ever.

And as their standing in righteousness and truth is to be indefectible, indefectibility attaches also to the place of supremacy assigned to them in the earth. "To thee shall it come, even the first dominion; the kingdom shall come to the daughter of Jerusalem." (Micah iv. 8.) "The nation and kingdom that will not serve thee shall perish; yea, those kingdoms shall be utterly wasted." Such supremacy indeed is the necessary result of having IMMANUEL for their king. He it is that "shall have dominion from sea to sea, and from the river unto the ends of the earth. They that dwell in the wilderness shall bow before him; and his enemies shall lick the dust. The kings of Tarshish and of the isles shall bring presents: the kings of Sheba and Seba shall offer gifts. Yea, all kings shall fall down before him: all nations shall serve him. . . . . . His name shall endure for ever: his name shall be continued as long as the sun: and men shall be blessed in him: all nations shall call him blessed." (Ps. lxxii. 8—17.)

Such is the promised heritage of *Israel.* Know
ing the tendencies of man's nature, we cannot wonder that Gentile Christendom should have coveted
that inheritance. "Wise in their own conceits,'
Gentile Christians early began to boast themselves
against Israel; to speak of them as having forfeited their ancient blessings, and to arrogate to
themselves glories that are reserved *for Israel and
Jerusalem alone.*

When, in the days of Constantine, "Christianity
ascended the throne of the Cæsars," the true doctrine of the Millennium received its death-blow.
How could that be waited for that had already
come? Was there not a *present* King, *present*
exaltation, *present* glory? The Dragon *was* van-
quished; Christ *had* triumphed. Zion *was* established. Such was the prevailing sentiment of the
day.

"The history," says a recent writer, "of this age (the
fourth century) opens badly. Eusebius is a semi-Arian, a
rationalist, and almost a præterist. In prophecy he can
believe nothing but what he sees; he must be shown a
fulfilment either in history or in passing events. . . . .
At first he received the Apocalypse: since he explained
the seven seals as the obscurities of the Old Testament
prophecies, which Christ, the Lion of the tribe of Judah,
now opens. Afterwards, the establishment of Christianity
seemed to interfere with the literal millennium; and then
Eusebius *must needs reject both Apocalypse and Millennium.*"

"Lastly, Constantine begins to rebuild Jerusalem as a
Christian city: and now the Apocalypse seems to be thought

of again, for this may be, perhaps, *the beginning of the new Jerusalem.*"*—Maitland's "Apostolic School of Prophetic Interpretation."

---

* The words of Eusebius are : " It may be that this was that second and new Jerusalem spoken of in the predictions of the Prophets, concerning which such abundant testimony is given in the divinely-inspired records." (See Eusebius, " Life of Constantine," lib. iii. ch. 33.)

Constantine's words were, " Now that the Dragon is removed from the administration of affairs, through the providence of the supreme God, and by my instrumentality, I imagine that the divine power has been made clear to all men." (" Vita Constantini," lib. ii. ch. 46.)

Such being the temper of the day, we cannot wonder that Jerome and others who had been trained in this atmosphere should, in expounding such a Psalm as the lxxii., comment thus : " He shall come down like rain upon the mown grass [or according to Jerome " fleece "] as showers that water the earth." Jerome understands this of the incarnation, for like as " when the rain descended on the fleece of Gideon when the earth failed from drought, so He [the Son] passed into the Virgin's womb by the infusion of the Holy Ghost. For the earth of the human body was dry when He came, and was unable to bring forth any fruit of holiness, but He, when He had given it fecundity by the distillings of His own preached Word, filled it with fruit for evermore."

We read in Is. xxvi. and xxvii. that *after* the Lord shall have come out of His place to punish the inhabitants of the earth for their iniquity (see Is. xxvi. 21, and compare it with Rev. xix. 11), and after He shall have "slain the Dragon that is in the sea (see Is. xxvii. 1, and compare it with Rev. xx. 1), and after the mystical body of Christ now dead shall have arisen (see Is. xxvi. 19, and compare it

The halo of deceiving brightness in which Eusebius and Constantine had contrived to invest the period in which they lived was soon rudely dissipated. Heresy, strife, and persecution, distracted the Church; bloody and destructive wars devastated the world; and civilization seemed likely to be swept from the earth by the fierce inroad of bar-

---

with Rev. xx. 4), that then, not before, "Israel shall blossom, and bud, and fill the face of the world with fruit." (See Is. xxvii.) The earth is replete with "thorns" *now*: Jerome and Christendom would have us think that it is filled with the myrtle and the rose. No trifling difference this. The thoughts of men, and the thoughts of God are separated as pole from pole.

On the words, "*in His days shall the righteous flourish, and abundance of peace as long as the moon endureth,*" Jerome observes that this "was fulfilled in the days of Augustus Cæsar when the Lord was born (cum Dominus de thalamo virginale processit) for that, at that time all wars ceased in such a manner that that also was fulfilled which another Prophet sang, saying, "They have changed their swords into ploughshares, and their spears into pruning hooks." On the sixteenth verse he observes, that "above all the heights of secular greatness, the grace of baptism, whereby sins are taken away, is exalted: and that, therefore, the saints born again by baptism, shall flourish from the heavenly Jerusalem, shining forth in the adornment of their good works (bonorum operum ornamento splendidi)." The words, "all the earth was filled with His glory," were fulfilled when, as we read in the Acts, "the Holy Ghost was poured out on all flesh." Jerome, therefore, could have found little difficulty in assenting to all that Constantine and Eusebius had said.

baric nations. In a little more than a century after the death of Constantine, barbarian hordes dominated over Western Europe. The words, *Væ pregnantibus et nutrientibus* (Woe to them that are with child, and to them that give suck) were so constantly in the lips of Jerome that they were said to be his motto: and yet, although the whole earth groaned and was filled with violence, it was still taught, that the saints were reigning, and that the millennial promises were being fulfilled!! If war meant peace; swords, ploughshares; and anguish, joy, it might have been true.

It is not to be denied that out of this vortex of universal woe, something that called itself Christianity contrived to raise its head. About the year 448, whilst Northern Africa was being occupied by the Vandals, and whilst Attila and the Huns were subduing Gaul and Northern Italy, Leo conceived the hope of making Rome the head and centre of the *Christian* world. "The project (says Maitland) was to be carried out by means of two assumptions: the first, that St. Peter had received the right of exercising dominion over the other Apostles: and the second, that he had been made Bishop of Rome, and had bequeathed to the Church of that city the inheritance of his supposed powers."

In a sermon preached on the festival of Sts. Peter and Paul, Leo thus expressed himself respecting Rome:

"These are the men that have promoted thee to so great glory, that being a holy nation, a chosen people, a royal and priestly city, and made head of the world through the holy seat of the blessed Peter, thou shouldest rule more widely through divine religion, than before by earthly sway. For great as thou didst become by thy many victories, extending by land and by sea thine imperial rule, the toils of war have won thee less than what Christian peace has placed beneath thy feet."

"This sermon of Leo became in turn a text, upon which his successors loved to dilate. Clement the Eleventh, preaching in 1706, thus hails the fulfilment of Leo's hopes:

"Rome exults in the most firm foundation of the apostolic rock: so raised to the summit of human affairs, that she now rules more widely through divine religion, than before by earthly sway. . . . Henceforth thou shalt be called the city of the Just One, the faithful city, the new Jerusalem; even the same that John saw coming down from heaven, prepared by God as a bride adorned for her husband; by imitation of which other things become fair, by comparison with which, foul. Hear this, you that inhabit the city of the Holy One, the city of the Just One, the faithful city, the new Jerusalem, by imitation of which other things become fair, by comparison with which, foul. It is monstrous to be in Rome, and not to be holy."

Whether they who first put forward these proud pretensions really believed their claim to be valid, we will not now enquire. It is very evident that some such conclusion as that which they drew must be deduced from the premises supplied by the instructors of Christendom, who, wellnigh with one voice, affirm that the time for the promised supremacy of Zion and the reign of the saints is come. The Catechism of Pope Pius IV. is sufficient to show the use that has been made of the deceiving notion. Grant its premises, and we cannot escape the conclusion. Nor can we deny, that if there be such a thing as Revealed Truth, it is utterly impossible that the world should be blessed until that Truth be supreme. *That*, Reason teaches us. Accordingly, Scripture declares that there is to be a time when Truth *shall be* supreme. If, therefore, Satan can deceive any into the belief that the time for the supremacy of Truth *is come*, and if he can also induce them to believe that *they* are the constituted guardians and administrators of that Truth, such men will certainly deem themselves to be the appointed regenerators of the world, and the sure healers of all its woe. They who are once deceived into this notion, are under a spell that no human power can break. Superstition, when it combines itself with ambition, is a terrible force. When strong natural tendencies are stimulated by supposed religious duty, the effect is potent, and the consequences disastrous beyond conception. One spark of deadly falsehood from the pit can kindle a conflagration that may extend to the world's end.

That this delusion has, since the Anglo-Catholic party appeared, been working potently in England is obvious to all who do not, ostrich-like, bury their eyes, and refuse to see. It is true that many, perhaps the most influential of that party, do not seem inclined to follow fully either Constantine or Leo. They would go back (not, indeed, to the Apostles, but) to the days BEFORE Constantine, and would found, on other principles than his, and on a basis wider than that which Vaticanism sanctions, a scheme for the union of Apostolic Christendom. Political Vaticanism they abjure; and by the vehemence with which they sometimes assail it, they deceive many into the belief that they are the enemies of *doctrinal* Popery; though *doctrinal* Popery they adore. Nor would they refuse, if Christendom were doctrinally one, that it should have a visible symbol of its unity in the precedency (under certain restrictions) of the Chair of Peter. They seem to think that Rome has narrowed her basis too much to be genuinely Catholic. They desire not so much uniformity as unity; and say that he who sits *as a Priest* upon his throne, destroys the distinctive characteristic of his supremacy, if, forgetting that his authority is wholly spiritual, he condescend to place himself on the level of earthly monarchs by wearing a temporal diadem. Secularity, in certain aspects, they deem to be Babylonianism. Yet *Catholic doctrine*, and *Catholic dogma*, they love with an intensity that Vaticanism cannot equal. Their doctrine of Development enables them to be eclectic; so that they can select some, and reject other, of the things which Infallibility has in past ages sanctioned; for development implies progress, and progress implies alteration; so that that which may be quite right in the youth-time of Catholicism may be quite wrong in its more mature age. The text, "His [Jehovah's] foundation is in the holy mountains," may be regarded as their motto. Churches whose Priesthood can claim for itself apostolic lineage have their foundations in those "holy mountains"; and such Churches, therefore, are or should be the centre of Jehovah's government in the earth. To *them* the promises made to Zion belong. *Their* standing is indefectible; and to them all nations shall finally bow. He, therefore, who resists the establishment of Catholic unity as the only right centre of the earth's government, resists God. Such is a brief, but I believe not untrue outline of the system that has for many years been working potently in England.

It may seem strange that men belonging to this High School—men who would jeopardise every thing for the maintenance of Catholic Dogma; who hate Radicalism, and loathe and despise Dissent; who believe (and in this they are not wrong) that all power, both in the Church and in the world, comes from God; who scorn the right of private judgment, and consign all who are not brought into their apostolic circle, to the uncovenanted mercies of God—it may seem strange that such men should consent to enter into close political alliance with those who repudiate all dogma, and assert that governments as governments should be atheistic, and maintain that if a man be sincere, it matters little (for purposes of legislation at any rate) whether he be a Catholic or a Protestant, a Deist or a Blasphemer. It seems strange that men who have drunk in from their earliest years the doctrines of Oxford Catholicism, should ally themselves to Destructives, and help the efforts of men who, if they succeed, would root up all ancient land-marks, resolve Society into its elements, and make "men like the fishes of the sea, like creeping things that have no ruler over them."

Do, then, these votaries of Neo-Catholicism indeed love chaos, and hate order? No. They hate chaos, and love order; that is to say, their *own* order—the order that *they* approve. The foundation of Jehovah is in the "holy mountains." All order, therefore, that hinders and resists the establishment of *their* "holy mountains," they abhor. Consequently, the present order of society, which does effectually hinder it, they abhor.* They wish

* Protestant ascendency in Ireland hinders it; therefore the Protestant Church was disestablished there; and seeing that the system of Land-tenure in Ireland is a remnant (the last remnant nearly) of Protestant ascendency, that also should be abolished. Protestant ascendency is regarded as the great Upas tree—to be destroyed wherever it exists.

Mr. Froude, the historian, will not be suspected of any peculiar attachment to Protestantism in any of its forms,

to destroy it that they may make way for a better order—the order of God's Church—the order of the true Zion—the order to which the promise of indefectible establishment is made. Empires may

---

yet even he, in a recent article on the Irish-land Question writes thus :—

"But there is another and very serious question. What did Mr. Gladstone mean by sanctioning this Act of his Irish Secretary? Mr. Gladstone does not know Ireland well, nor its history well; but he has attended to both, he has formed views about both, and to some extent must have understood what he was doing. It may have been that he was merely careless, that he wished to please his Irish supporters, to pass pleasantly through the remainder of the Session, and to save himself from being troubled, for a few months at any rate, with Irish disturbances. But Mr. Gladstone is not a person to act in so serious a matter without a clearer purpose, and expressions have dropped from him which betray a feeling of another character. The landowners were a branch of the Upas tree, a surviving symbol of Protestant ascendency."

I am not regarding this question politically. I am not defending either the Irish Church, or anything else in Ireland. But I cannot close my eyes to the fact, that they who have chiefly contributed to the overthrow of the Established Church in Ireland, attacked it not because of the evil that was in it, but attacked it because they hated Protestant Truth. Shall the servants of Christ for selfish purposes, or for any other reasons, co-operate with the haters of God's Truth? If they do, God will not allow them to wield His *spiritual* weapons. They will drop out of their hands. They will find themselves partakers of other men's sins, and will, unless they repent, be smitten with them.

crumble; nations fall; democratic fury may rage, and possibly for a season subvert society. Order may temporarily give place to ruin; yet we, or if not we, our principles, seeing that they belong to "the holy mountains," will outlive every storm; subdue every enemy; come, like "the woman clothed with the sun," out of the wilderness into which Antichristianism may drive us for a season, and so become the new and lasting centre of order and stability to the world. Surely, every thing should be swept away that hinders the attainment of an end such as this. Soldiers who are preparing for the defence of a fortress that is about to be besieged, are not accustomed to spare the city that may surround their citadel. They sweep it all away, lest it should supply shelter to the foe, or narrow the scope of their own operations. Destructiveness becomes, under certain circumstances, a virtue. It may be needful to destroy before we can construct. Whilst destroying, we welcome only those who will help us to destroy: when constructing, we repudiate all who will not help us to construct. The allies of to-day may be cut off to-morrow. Versatility and change are necessary concomitants of *development*.

Many, no doubt, will scoff at all this. They will say, "Such notions, if they anywhere exist, are the wild dreams of fanatics, and will perish under the weight of their own folly." That may be so; yet experience teaches us that energetic fanaticism is not to be despised. It may originate in the cell of the recluse; but the seeds there prepared may be scattered over all nations. Peter the Hermit was a fanatic, and so was Ignatius Loyola; yet their fanaticism shook the world: and the principles of the latter are in operation still, as the condition of Germany, of France, and of England, shows. Any system in which conscience, reason and principle, are ruthlessly subjugated to an authority so omnipotent that it can change the nature of things, making virtue vice, and vice virtue, is sure, more or less, to prosper in a world like this, where, for the most part, expediency and self-interest rule. Cold, calculating, ambitious, reckless hearts (like that of *Warenne* in the tale of Father Clement, and there are many such) love such systems, for they subserve their ambition and their selfish interests: whilst others, weak, timorous, amiable and morbidly sensitive (like *Dormer*) are awe-stricken, and bow to them in superstitious fear. The frivolous too, and the sentimental, and the imaginative, become ready victims. Moreover, there is a mighty one ever near us, who well understands the character and ways of man, and who can, when God permits, so touch the inward strings of thought and feeling as to make the strongest, and the wisest of men, slaves to superstition. Was not Charles V. an example? Fanaticism, therefore, is not to be despised. Jesuitism in its origin, must surely be regarded as the offspring of fanaticism. Yet, has not Jesuitism lately, by its alliance with Socialism and the like, made Germany tremble? Does Prince Bismarck think that fanaticism is harmless? In England the coalition that has been formed between Anglo-Catholicism and the extremes of Radicalism and political Dissent, has resulted in placing in India a Romanist Viceroy, and bringing into our legislature at home a declared *Atheist*. Would these, and like things—things that have caused, and do cause, the hearts of those who fear God to tremble—would these deeds have been perpetrated without the aid and the countenance of Anglo-Catholicism? That there will finally be a conflict between Anglo-Catholicism and those whom it is now using as its tools, is certain: but such a conflict is not to be lightly thought of. It may shake Europe to its centre. In some way or other, fearful chastisement must come: for our carelessness, ignorance, and worldliness are earning it.

The energy and the daring that have recently characterised some of the movements of Anglo-Catholicism in England seem to indicate that they who have inaugurated or led these movements, regard themselves as being in undisputed possession of the field. Doubtless, the circumstances of the times greatly favour their plans. A lax, careless indifferentism to religious truth, and a devotion to the "material interests" of men, have leavened vast masses. Among such it is very certain that none will be found to do battle for Truth, or to make for it any sacrifices. The philanthropy of the day is, for the most part, unguided by religious knowledge, and is soon conciliated by a sop, or a smile. Political Dissent, instead of restraining, stimulates aggression on any thing to which it is ecclesiastically or secularly opposed. Moreover, there is in men's hearts an innate love of change that naturally inclines them to side with those whose object it is to alter, or to destroy. Firmness of action too, and stedfastness of purpose (qualities peculiarly found in men who serve a fixed dogmatic system) contribute greatly to success in the midst of a multitude that have few definite thoughts, or if they have them, have no steady principle of cohesion. It is not, therefore, to be wondered at, that men who can deceive themselves into the belief that the strength of "the holy hills" is indefectibly theirs, should in the present social and political condition of England, be daring in excess. Even if all the winds of heaven were loosed from their prison-house they would not be disconcerted, because they imagine that they have the power to control the storm, and to hush it as soon as it has wrought the effects which they desire.

In "*The Times*" of June 23rd, 1880, Mr. Gladstone, speaking on the Bradlaugh case, is reported thus :—

"They are about to take up the position of objecting on religious grounds to the appearance of Mr. Bradlaugh in this House. Very nearly two centuries ago this House was the scene of conflicts of the same kind as the present. Those conflicts lasted from the beginning of the last century down to the year 1828. They were really carried on in defence of the Church constitution *de jure* existing. Up to that time it was assumed that members had qualified themselves by a solemn act of communion with the Church of England.* The House was rallied by that call. The doctrine was perfectly clear and consistent. But in that

---

* This is a mistake. It was never required of Members of Parliament that they should belong to the Established Church. The Nonconformists were not excluded from Parliament by the Test and Corporation Acts. The Romanists for special reasons (reasons justly held to be valid by the Nonconformists in the reign of James II., and by Wesley afterwards) were excluded.

It is not the desire of God that a fictitious position should be assumed by any. For those to pretend to belong to the Church of Christ who have not by personal conviction, and personal faith, been brought into it, is untruthfulness; and God cannot sanction untruthfulness. The world is one circle; the Church another. These circles are not to be confounded. But the world, though it is not the Church, has its duties. Men, as men, have certain relations to God as their Creator, which He expects them to observe: and the secular governor is bound to do all in his power to prevent men from violating these natural relations. On the other hand, the servants of Christ are to act as Paul did towards Felix and Agrippa. He did not seek to force them to any thing, nor did he seek to enter into any alliance with them. If he had, he would have sinned like Jehoshaphat. All that he did was to reason with them out of the Scripture, and to impart to them all the light they were able to receive. He attempted nothing more. He left the rest with God.

year the Test and Corporation Act was repealed. That was followed by another rally. The country was next told to rally round the Protestant constitution of the House. That controversy was mixed up to a certain extent with the preceding one. Then it was felt that that was not a satisfactory solution; and, in consequence, Roman Catholics were admitted to take seats in the House, many years having been angrily spent in the contest. (Loud cheers.) Then, having got rid of the Protestant character of the House, the country thought there could be no further religious war. But then a fresh cry arose. That cry was on the Christian character of the House. It was observed and insisted upon that nothing could be so injurious, so destructive to the religious character of the country, as the admission of Jews. For thirty years that controversy raged. There was another rallying cry to adhere to the Christian character of the constitution. That era was closed by Jewish emancipation, if, indeed, it can even now be said to be absolutely closed, for we have found by this amendment that there is another branch of the same contest, and it is sought to bring within the discretion of the House the question whether this or that member shall be allowed to take his seat. It was justly and truly said last night by my right hon. friend [John Bright] that the House was now invited to make a final rally for the theistic constitution of the House. We have been driven from the Church ground; we have been driven from the Protestant ground; we have been driven from the Christian ground; and now it appears there is to be a final rally upon this narrow and illogical basis of theism. (Cheers and counter-cheers.) That will go whither your Protestantism and your Christianity have gone. . . . There is a theory that it does not matter what god you worship provided you worship some god or other; and I must confess that in my opinion there is greater danger of irreverence and impiety in this kind of loose and rambling theory than there is in a frank acknowledgment of the absolute separation that has been drawn in the spirit of the law of this land, and, as I believe, in the letter of the law of this land, between civil duty and religious duty."

And contrast now this speech with the words of Mr. Gladstone in 1838:—

"The point of view from which it is now proposed to contemplate and discuss the question, is that which men occupy as members of a State; and the aim is to show, that the highest duty and highest interest of a body politic alike tend to place it in close relations of co-operation with the Church of Christ. It is from this position that I propose to regard it; first, because the combatant in defensive warfare naturally resorts ἐπὶ τὸ κάμνον, to the quarter which is threatened and in danger: because the Church is not likely to be the moving party in measures for the dissolution of this connection, while the State has, it is too certain, given signs, though perhaps unconsciously, of that inclination; and therefore it is the mind of the State, not of the Church, which requires to be more fully exercised upon this subject, in order to the better knowledge and fulfilment of its duty.

"But besides the fact that we are more ignorant of our duty as citizens than as churchmen, in respect of the connection, we shall find another reason for instituting the investigation in the former capacity rather than the latter. The union is to the Church a matter of secondary importance. *Her* foundations are on the holy hills. Her charter is legibly divine. She, if she should be excluded from the precinct of government, may still fulfil all her functions, and carry them out to perfection. Her condition would be anything rather than pitiable, should she once more occupy the position which she held before the reign of Constantine.

But the State, in rejecting her, would actively violate its most solemn duty, and would, if the theory of the connection be sound, entail upon itself a curse. We know of no effectual preservative principle except religion; nor of any permanent, secure, and authenticated religion but in the Church. The State, then, if she allows false opinions to overrun and bewilder her, and under their influence, separates from the Church, will be guilty of an obstinate refusal of truth and light, which is the heaviest sin of man. It is of more importance, therefore, for our interests as a nation, that we should sift this matter to the bottom, than for our interests as a Church. Besides all which, it may be shown that the principles, upon which alone the connection can be disavowed, tend intrinsically and directly to disorganisation, inasmuch as they place government itself upon a false foundation."—*The State in its relations with the Church*, *p.* 3, 4.

"The principles," says Mr. Gladstone, "upon which alone the connection can be disavowed, tend intrinsically and directly to DISORGANISATION." Therefore, in his speech, according to his own acknowledgment, he has sanctioned principles which "tend intrinsically and directly to DISORGANISATION."

It will be said by some that Mr. Gladstone's principles have changed since he wrote these words. Where is there any evidence of this? Has he ever said so? All the facts of his past and present history seem to show that he has acted, and is acting, in strict consistency with his principles. His contention is, as I understand it, this. The State is bent on disorganising itself. He and others have warned it; but in vain. It is now best that the disorganisation should proceed. The time will come when re-organisation will be sought after. "She whose foundations are in the holy hills"* has alone power to effect this re-organisation. "*Her* charter is legibly divine." She alone can mould rightly human things. If she fail in doing this, there is nothing else that can succeed. If such be his views, I can see no inconsistency in his past or present course. I can see no evidence that he has changed his views. He has tried many parties, and many individuals, but will use only those who are subservient to his views. It would be strange indeed that his mind (surely not one wanting in perspicacity) should in the year 1838 have discerned "disorganisation," whilst comparatively it was yet in the distance, and that it should not recognise it in 1880 when it is well-nigh present.

In one of the Nonconformist journals—"The

* "*Her foundations* are in the holy hills" is a misquotation. The words of Scripture are, "*HIS* [*i.e.* Jehovah's] foundations are in the holy mountains." In the next Dispensation, when the present age of evil shall have passed, and when the Son of Man shall have been brought before the Ancient of days, and shall have taken to Himself His great power (see Rev. xi.), the foundation of the Divine Government in the earth will be in the holy mountains of Immanuel's Land, and more especially in Zion. The Church will not rule as the vice-gerent of Christ. He will Himself reign.

Methodist"—there is the following comment on Mr. Gladstone's speech. The italics are mine.

"Very important principles are involved in the Bradlaugh case. It is to be regretted that these principles cannot be separated from such a man. There is hardly a member on either side of the House who is not disgusted that this discussion has arisen in connection with such a name. To the reverent mind of the Premier the entire episode must be painful. But what can be done? There are two courses open. One is that civil right and orthodoxy should coincide. This course is impossible. The nation has withdrawn from it. *It can no more turn back than the river can turn back.* The other course is *to accept the absolute separation between civil duty and religious belief.* This is the logical issue which must come, as Mr. Gladstone put it in his magnificent speech on Tuesday night. The House has already lost its Church, its Protestant and its Christian character, and its theistic character will have to go. We have admitted Nonconformists, Catholics, and Jews, and we shall have to admit Agnostics and Secularists. Illogical compromises last only for a time. Practical difficulties must be dealt with."—"The Methodist," June 25, 1880.

It will be an awful thing for those who have thus abandoned God, to meet the "practical difficulties" which must and will arise. They little know what those "practical difficulties" will be, and what the Scriptures have revealed respecting them. We are told that we must quietly "*accept the absolute separation between civil duty and religious belief.*" We are to believe that men who avowedly disown God their Creator may as efficiently discharge the duties of life as those who acknowledge Him. Does the history of the past, either in the case of nations, or individuals, teach me that this is so? Does Holy Scripture teach me that it is so? Does not the Scripture say that they who refuse to retain God in their knowledge are by Him given up "to a reprobate mind to work all uncleanness with greediness?" (See Rom. i. 28.) Are men of reprobate minds as fitted to discharge their duties as men who are not thus hardened? The light and the power necessary to fulfil "duties" must be absent where the one source of light and of power is withdrawn from. The lessons both of Providence and of Revelation teach me that curse must rest on the head of those whom God abandons. Moreover, I should have thought, that if wrong steps have been taken, it would be our duty to humble ourselves and to confess: and if they cannot be retraced, to pray earnestly that we may be preserved from advancing further in the path of evil. But no: instead of this, we are exhorted to add sin to sin, and boldly to advance in the course on which we have entered. To confess, and to repent, would be weakness and folly, and would compromise our consistency. Consistency is one of the chief of virtues; our consistency, therefore, must be maintained, even if the result be the abandonment of God, and the refusal to own any unseen Being superior to ourselves. A new path this, which nations have not hitherto trodden. The Scripture, however, tells us that in the latter times a large section of mankind will persist in treading it; and it reveals to us the end.

If any say, We desire that the disease should be developed in order that the remedy may be applied, we reply, Does God command us to encourage rebellion against Him that we might thereby reach good? Even if a remedy were beyond, we are not to reach it by an increase of sinning. But there is no remedy beyond. The expectation is delusion. The wickedness and falsehoods of the so-called Church have brought on the plague. The votaries of a false Christianity may stimulate that plague; but they are powerless to heal it. Jehovah hath not yet "founded Zion." His foundation is not yet in the holy mountains. (See Is. xiv. 32, and Psalm lxxxvii.) They who pass deliberately over "the narrow ledge of Theism" will find themselves plunged into a gulf of hopeless Diabolism. They will have said, "Evil, be thou my good," and with Evil and the Evil One they will dwell. The Atheistic blasphemy that is to reign over all the Ten Kingdoms of the Roman World will finally be abolished; but not by any human power. Christ will meet it by the manifestation of Himself in glory. Then all they that have abandoned God, and all the votaries of a false Catholicism, who have blocked the way of Truth and stumbled men by their corruptions, will (unless they repent) find themselves swept, together with their victims, into that outer darkness where hope never cometh. "THE Apostasy" is fast coming on: but it will be met by no remedy. Destruction is not remedy. When the blood of the Lamb of God is trampled under foot, what can remain except a "fearful looking for of judgment and fiery indignation, which shall devour the adversaries."

CHAPTER XV.

PRESENT TENDENCIES.

§ I.

It is not to be wondered at that they who believe that there is now existent in the earth an organised Body "dowered" by Christ with indefectible grace and gifts and power, and appointed by Him to rule supremely in the earth—it cannot be wondered at that such men should be utterly dissatisfied with every relation which that body now stands in, or indeed ever has stood in, towards the secular Governments. They would have been satisfied if the Governments had unreservedly bowed to the Church, and owned her sole supremacy; but in default of this, they would prefer that the State should avowedly renounce all regard to religious Truth, and abandon even "the narrow ledge of Theism." Thus would be *proved* not only its incapacity for *ruling* the Church, but also its utter unfitness for any kind of association with it; for how can they who spurn God have any connection with the inheritance of God? How could the regulation of the moral and educational, and religious arrangements of Society be entrusted to a Body that included within itself blasphemers of God and that might become *wholly* atheistic? By isolating the Church therefore, and bringing it back into the condition that it occupied before the days of Constantine, they hope that it will be once more placed in a condition to assert successfully its claim to be the moral Ruler of the earth—a claim more likely to be recognised when men shall have been made sensible of the effects of the "disorganisation" and "curse" (to borrow Mr. Gladstone's expressions) which the rejection of God and of the Church will have entailed. High expectations these. If indeed it were true that the present professing Gentile body inherited the earthly promises made to Israel—if they were appointed to reign, and not to suffer—if restoration to the position which the professing Church held before the days of Constantine were a restoration to the position held by it while the Apostles lived; if Ecclesiasticised Christianity were Apostolic Christianity—if the conditions implied by all these momentous "*ifs*" were duly met, then, no doubt, the fulfilment of these lofty anticipations might be looked for. "The holy mountains" would be established; Truth would reign; Kings would be its "nursing fathers," and their queens its nursing mothers."

Few who read these pages will need to be told that the conditions have *not* been fulfilled, and that the expectations will *not* be realised. They will be as the dream of a night vision. Nevertheless, the effort to maintain these principles will doubtless bring to many eternal ruin, and may cause vast national calamities. Nor is this the only system by which we are threatened. Another "school of thought" has arisen in England (which also has been chiefly nurtured in Oxford) whose principles, though operating more slowly, are more suited to the temper of the day, and more akin to genuine Antichristianism. Men of this School bow to no prescriptive authority. They decline the control of Scripture, and of Tradition, and of everything dogmatic and fixed. Refusing to admit the absolute truthfulness of Scripture, they think apart from it. They little trouble themselves to enquire whether there be, or be not, a millennium—whether any promises yet unfulfilled, have, or have not been made to Israel—whether there has been, or has not been an Antichrist. They prefer to leave such questions to unpractical minds who are fond of amusing themselves with phantasies. One thing (say they) is obvious. The world must be governed. Society must not be allowed to become disorganised. If it cannot be controlled by old and effete principles, we must devise new. Channels must be made in which the thoughts of those who profess, and of those who renounce, a regard to religious truth may be made to flow peaceably together. Influence is the great thing needed. Whatever promotes such influence is to be valued—whatever promotes it not, is to be eschewed. The world is not now what it was three centuries ago. Knowledge and civilisation were then confined to a corner of Western Europe. Now they are extended over the globe. Old countries are being revived: new countries are being penetrated. A world-wide sphere requires world-wide principles. Consequently, the narrowness of former ages must be abandoned; and influence be sought by devising some system that may adapt itself to, and become the expression of, the thoughts of civilised humanity wherever found.

It was to be expected that men of sagacious mind, indisposed to submit to the lordly restrictiveness of Anglo-Catholicism, and yet anxious to secure the influence that Religiousness bestows, would look upon a re-organised State-Church as the instrument best suited to forward their designs. Men, *for the most part*, are prone to worship something. They are not disposed to renounce every form of religiousness, but they like their religiousness to be dignified, and to be æsthetically, if not intellectually, pleasing. A Catholicism founded on dogma would not suit them; but a Catholicism founded on a well-nigh unrestricted comprehensiveness they welcome. A Church united to a latitudinarianised State must latitudinarianise itself also. If the State adapts itself to the mind of Society, and becomes expressive of its will, the Church must do likewise. A latitudinarianised State is not devoid of order: a latitudinarianised Church may have order too—an order, however, which expediency, not principle, must supply; for fixity of principle would lead to dogma, and dogma is incompatible with comprehensiveness, and comprehensiveness cannot be dispensed with, for Society (which refuses to be regulated except in conformity with its own will) demands it. Therefore, unless Society be left unregulated, its will must be obeyed. There must be *self-government*. That the liberty of human thought and action should be restrained only by such checks as men may agree to impose one on the other, may be a plan well-pleasing to *them;* but it brings into necessary independence of God. The service of *His* people is altogether guided by dogma—dogma founded, not on the Traditions of men, but upon His written Word. When, therefore, *dogma* is repudiated, God is abandoned, for God controls by dogma. By Revealed Truth God authoritatively moulds the minds of those who obey Him. If men refuse to bow to His authority, and persist in being the guides of their own steps, there can be only one result. They must become the slaves of Satan, for he it is who "worketh in all the children of disobedience."

Reflections, however, such as these, have not the slightest weight with those who are bent on having the mind of Society expressed, *secularly* by the State—religiously by an all-co— prehensive Church. As to the definition of the term "religious," that is a *point* which men are to determine for themselves. They who can be induced to believe that the inspiration of Joshua, and of David, and of Solomon was *in kind* not different from the inspiration of the chieftains, poets and philosophers of the world, will readily admit, that to serve men *secularly*, is to serve God *religiously*. The cultivation of philosophy, science and art, is to be recognised as the fulfilment of a *religious* duty. He who studies Shakespeare is to be regarded as studying *a* book of God. Schools of philosophy and art are places where, as truly as elsewhere, we may serve and worship God. Such are the dogmas of this system.

At present, however, Society in England, though grievously perverted, is not sufficiently debased to receive these principles in their entirety. It has been suggested by some that Westminster Abbey, and like structures, might be used as places where successively Philosophers might lecture neologically, and Cardinals celebrate High Mass, and Ritualists practise histrionically, and Protestant Evangelists preach; yet even this comparatively moderate suggestion has met with little favour.* Another plan,

---

* A proposal of this kind lately made by Dean Stanley has thus been commented on in one of the London jour-

therefore, having a more religious aspect, is proposed. Yet it points in the same direction.

---

nals. Christians would do well to note the passage I have placed in italics.

"The recent letter of Dean Stanley advocating a peculiar form of ecclesiastical unity was adverted to at the Congress, but found no supporters. The idea of the Dean is that even the present subscription, which requires from the candidate for ordination nothing but assent to 'the doctrine' of the Church, should not be required. He would also open the churches to the services and sermons of all Nonconformists. This is certainly a bold advance towards a very elastic and very Erastian Catholicity. The only part of the world where it is even partially realised is in China, where Buddhists, Taonists, and the disciples of Confucius share the same temples, mutually tolerating rival symbols and controversial inscriptions. The Dean would evidently transform the churches *into something like those monthly reviews, now so fashionable, which open their pages to all writers of all parties and creeds.* In London itself this toleration is personally carried out by habitual sermon-hunters, who run after various preachers, hear Cardinal Manning one Sunday, Mr. Voysey on a second, Canon Liddon on a third, filling up after vacancies with Mr. Spurgeon or Mr. Stopford Brooke. The Dean's idea would be to keep the congregation stationary, and let the pulpit door admit in turn a supply of rival preachers to meet the modern demand for variety of doctrine and thought. We doubt the practicability of the plan; if practicable, we should doubt its utility. Congregations thus assailed by controversial disputants—the doctrine of one Sunday denounced on the next—would be apt to take refuge in a mild scepticism, they would suspend their judgment and accept nothing until they heard the advocate on the opposite side. In no age, in no country, has a church or a creed been thus transformed or split up. Then the proposal of the Dean that clergymen should not be asked for any 'subscription' is hardly more felicitous."

The periodicals and journals of Christians are, in their way, becoming just as latitudinarian as journals conducted by others.

§ II.

THE judicial decisions, first in the Gorham case, secondly in that of the "Essays and Reviews," and lastly in the Bennett case, formally widened the basis of the Established Church of England, so as to include Evangelicalism, Neologianism, Infidelity, and doctrinal Popery. The national Church has thus been made to express, not indeed the mind of Society as a whole, but the mind of three great sections of Society. Articles of faith have been professedly retained, but the adoption of the principle of "*non-natural* interpretation" has practically cancelled them. Recently, the Archbishop of Canterbury in Oxford, with the Popery of Keble College on one hand, and the scepticism of Balliol on the other, pronounced his panegyric on the course which Oxford has pursued of "honouring honest conviction wherever found."* The Croydon Congress soon followed, where the Archbishop had the gratification of seeing his principles practically carried out by Evangelicanism, Neologianism, and Ritualism consenting to unite, and acknowledge each other as "different Schools of

* See this further considered in "Europe and the East," p. 88.

thought" co-existent in one Church. This was a momentous step onward in the path of ruin, and has doubtless sealed the doom of Evangelicalism as a system, for it is a step that has not been repented of, but has been acquiesced in and even vindicated. Soon afterwards, the Pan-Anglican Synod was convoked at Lambeth. There the Protestant *Episcopacy* of the world virtually acquiesced in the principle of a latitudinarianised fraternisation. Lambeth seems desirous of erecting itself a centre for all peoples, and is (as we are told) receiving advances even from the ancient Churches of the East. It would aim, apparently, at establishing under its nominal headship, an *undogmatic* Church. One of the chief organs of public opinion in England, distinguished by the keenness of its sagacity, observes that the comprehensiveness of the Archbishop's last charge is so great, that apparently none need to be excluded from it who do not voluntarily exclude themselves :* and another journal, scarcely less influential, observes :

* The following is an extract from "The Times." The italics are mine. "The Archbishop of Canterbury, in the charge he delivered yesterday at Croydon, goes over a wide ground. The Church of which he is the head has been extending its influence at home and abroad. It has established relations more or less intimate with all Christendom outside the Roman Catholic Communion. Lambeth has become the common centre to which a hundred and sixty-two Bishops are looking from all parts of the world; and there are many Christian bodies, too, not ruled by

"The Archbishop of Canterbury preaches in his visitation charges *a tolerant and thoughtful Theism,* Bishops, but which the Primate recognises as separated from his own Church by an indistinct and fading line. All these are dealt with in the charge. The claims of all of them are acknowledged, and a hearing is promised to the cry for help which is continually coming in from a good many of them. At home the Archbishop's charity is limited only by the perversity of its possible subjects. He will hold out his hand to everybody who will take it; and if he finds his hand refused, he will not the less cultivate such more distant friendly relations as may conduce to the general good. The Archbishop, in a word, finds himself the acknowledged head, not only of the Established Church of England, but of that larger and looser communion which was represented two years ago at the Pan-Anglican Synod, and he is willing to extend the line beyond even this. *A common Christianity, and a desire, or even a non-refusal, to be taken in, are the sole conditions he insists upon. Where these essentials are to be found, he will sink all minor differences.* The Church of Rome is the one body which he has no near hope of acting upon, or acting with. The American Church, the various Eastern Churches, the Old Catholics, the Lutheran Churches everywhere, the Protestants of France and Spain and Mexico, all come within the sphere of his sympathy, and he demands recognition for all of them and help where help is needed. His ideal Church is naturally his own Church, or, together with his own Church, the Churches which most closely resemble it. The picture which commends itself to him is drawn from real life, a little softened down in some of its details, and a little aided by a determination to look at things as they ought to be and as they may be, rather than exactly as they are."

and while exhorting his clergy to believe in the Atonement and in the inspiration of Scripture, advises them not to define too closely the character of the doctrines they teach. To avoid barren controversy, to cultivate charity, and to do practical good, are the notes of our religious teachers today. From the point of view of the theologian, this may seem a falling off; but practical men find plenty of compensation in these not unwelcome signs of the times."

These strictures cannot be deemed excessive. Indeed they are the strictures of a friend. If the principles of "the Charge" were carried out, it would certainly bring the Established Church, and all that shall come within the Lambeth circle, to that "narrow ledge of Theism," which the House of Commons (after having for some time occupied it) has recently renounced. If the Scripture be abandoned (and it certainly is abandoned, if its inspiration and the strict definiteness and truth of all that it reveals is denied) the abandonment of God will soon follow.

The struggle between the dogmatic Catholicism of the Vatican and the undogmatic and elastic Universalism of Lambeth may be severe and long, and may bring calamity and ruin upon thousands. Yet the conflict of these two great Ecclesiastical systems will only render it the more easy for the secular Governments to establish their claims to supreme religious control. Without conceding pre-eminence, much less supremacy, to any one ecclesiastical or religious body, the secular Governments will undertake the patronage and regulation of all, on terms more easy than even pliable Lambeth could propose. If the duty of submission to the State be recognised, and the struggle for religious *supremacy* renounced, no further concession will be required, except on one point; but that point the servants of Christ will regard as vital, though by the multitude it will be esteemed as nought. Aggressiveness will be forbidden; and yet aggressiveness is a necessary characteristic of all faithful soldiership for Christ. "Think not," said He, "that I am come to send peace on earth; I came not to send peace, but a sword." These words, however, will be cast aside, and the peace of concord sought after—concord in error. If that be acquiesced in, nothing more will be required. There will not only be equality of *protection*, but equality of *position, emolument,* and *honour*, in all that the State dispenses: whilst they who refuse to join the unholy combination will be trampled down. This is the end towards which we are advancing, and Lambeth is smoothing the way: but we must not expect to see this system perfected and presented in its full development, until the Ten Kingdoms, into which the Roman World is to be divided, shall have been formed and federally united. The religious equality so long coveted by many, will then be attained: but it will last but for a moment. It will be an equality purchased by the abandonment of God: and the scourge of God will come. The despotic might of Antichrist will force all who fall within the scope of his power (and all the Ten Kingdoms will) into a unity, the like to which the world has never yet seen. All who fall within the scope of his power "shall worship him, whose names have not been, from the foundation of the world, written in the book of life of the Lamb that hath been slain."

How earnestly, then, ought we to pray for ourselves, and for others, that we might be preserved from being connected directly, or indirectly, with the circle of Lambeth, or with Vaticanism, or with Anglo-Catholicism. Separation unto the Bible, and to the Bible alone, was never more needful than now.

Nothing can be more important than that we should clearly recognise the distinction that subsists between those who are brought by personal faith into the fold of Christ, and all who are in the world besides. In the case of all who are saved, there is a period in their lives when they become, more or less deeply exercised in their consciences because of their condition in relation to God. They become conscious of the sin that is *in* them, and of the sin that is *around* them. They become troubled; and their trouble never departs until they recognise that there is "a fountain opened for sin, and for all uncleanness" in the blood of the Lamb, and wash there. Faith is reliance—reliance on God as testifying the efficacy of the atoning work of Jesus. God "preaches peace by Jesus Christ." All who truly receive that testimony are brought within the circle of grace. They become children of God *by redemption*. They belong to the Church of God; for the Church of God is composed of those who are sanctified by faith. They stand in a new spiritual relation to God, being quickened by His Spirit.

But, because such relations to God as these belong only to believers, we must not on that account ignore the relations in which men as men stand to God. Adam when first created was not the son of God *by redemption*, but he was the son of God *by creation*. (See Luke iii. 38.) The Apostle Paul also, when speaking respecting God to the Athenians, says, "*We are also His offspring*." Nor has God left Himself without witness among His creatures. The works of creation testify of "His eternal power and Godhead." (See Rom. i. 20.) He "sends rain from heaven and fruitful seasons, filling their hearts with food and gladness," and thereby declares His kind consideration for their need. He sends providential deliverances also, as when "they cry unto the Lord in their distress, and He bringeth them unto the desired haven." He has sent also into the world His written Word, and preaches likewise the Gospel of

His grace through Christ. All these and like things are mercies directed towards men as men. Are they to be ignored? Are no responsibilities contracted by neglecting or despising such mercies as these?

There cannot be a greater mistake than to suppose that because men are naturally born in sin, and therefore in moral distance from God, it matters not whether that distance be, or be not, increased. There are *degrees* of darkness, as well as *degrees* of light. All disobedience is sin: deliberate rebellion is sin, and apostasy is sin, but they are different *gradations* of sin. We have to beseech men not to "add rebellion" to their sin, much less apostasy. A conscience open to the appeals of Truth is different from "a conscience seared with a hot iron." One great object to be kept in view in the education of children is, to prevent the voice of *conscience* being silenced.

Was there no difference in the condition of Nebuchadnezzar when "his heart was lifted up, and his mind hardened in pride," and afterwards when he had been smitten and owned the justice of the blow, and said, "Now I Nebuchadnezzar praise and extol and honour the king of heaven, all whose works are truth, and His ways judgment: and those who walk in pride He is able to abase"? Did Daniel speak foolishly when, after recounting to Belshazzar the history of his father's chastisement, he said, "And thou his son, O Belshazzar, hast not humbled thine heart, though thou knewest all this; but hast lifted up thyself against the Lord of heaven; ... and the God in whose hand thy breath is, and whose are all thy ways, hast thou not glorified." Did Daniel think it a matter of no moment whether Rulers reverenced God, or defied Him? Did he adopt the vaunted maxim of these modern days and say, that "civil duty is to be separated from religious truth"? When Paul said to Agrippa, "I think myself happy, King Agrippa, because I shall answer for myself this day before thee, &c. .... King Agrippa, believest thou the prophets? I know that thou believest." Did he see no difference between the ignorant indifference of Festus and the sensibility of Agrippa to the appeals of Truth? Is it a matter of no consequence whether they who govern, *own* God, or whether they *blaspheme* Him? Do rulers who "kiss," that is, "do obeisance to the Son," stand in the same relation to God as those who say of Jehovah and of Christ, "Let us break their bands asunder, and cast away their cords from us." The second Psalm is God's own answer to that question. We cannot blot that Psalm out of the Word of God. It remains a standing witness against all those who would separate "civil duty" from religious Truth. What can be more monstrous than to be asked to believe that it matters not whether the world be ruled by men who are "founded together [such is the expression*] *against* Jehovah and *against* His Christ," or by men who, like our own Edward VI., seek to rule in the fear of God. Do the past histories of Protestant England and Catholic Spain present no contrasts? Will England, now that she has begun to despise and trample down the Bible, be morally and socially what she was when she reverenced the Bible? Facts are already answering that question.

In defiance of every lesson supplied by Scripture, fact, and experience, and even by Nature itself, we are on every side met by the cry that "the Bible has no more to do with Legislature than it has with a mechanic in making a machine"; that Governments as Governments should be atheistic, and that "civil duty is to be separated from religious Truth." The political Nonconformists, as a body, have adopted these principles. They have certainly been the chief supporters of a movement that has resulted in placing a Papist as Governor over the chief dependency of the Crown, and in introducing an avowed Atheist into the Legislature. We are asked to believe that power (which is the gift of God alone) can be wielded rightly by men who deny that which the Devils own—the existence of a supreme

---

* "The kings of the earth set themselves, and the rulers are founded together (נוֹסְדוּ־יָחַד), against Jehovah, and against His Christ." (Ps. ii. 2.)

God. Do we see no sign of the times in these things—no token of the coming Apostasy?

The history of Nonconformity is sorrowfully instructive. When the progress of the Reformation in England was arrested by Elizabeth and her successors, and when the Bishops, acquiescing in the requirements of the Crown, consented to retain and to sanction certain vestments, observances, and doctrines, in which the plague-spot of Popery still remained, and which should (like Amalek of old) have been unsparingly extirpated, the Nonconformists refused to yield. They manfully withstood the current, and refused the unholy compromise into which others entered. Accordingly, in their ranks, for many long years, the chief defenders of God's Truth were found. Although persecuted, even unto death, they "endured as seeing Him who was invisible." "They took joyfully the spoiling of their goods, knowing that in heaven they had a better and an enduring substance." The Established Church of this country has a terrible account to render touching these things. The ejection of 2,000 godly ministers by the Act of Uniformity, is a sin that has not been repented of. Its effects we shall never, whilst in this world, duly estimate.

When James the Second ascended the Throne, the days of the Established Church seemed numbered. He was avowedly a Papist, and his resolve was (just as it is the aim of certain Statesmen now) to seduce, or else to force the Established

Church back into Popery; or (failing that) to destroy it and to establish Catholicism on its ruins. James expected to find in the Nonconformists willing allies. He knew how grievously they had suffered, and he thought that by freeing them from those sufferings, he might bribe them to become his tools. He condoled with them therefore; professed to be the upholder of the principles of "civil and religious liberty;" repealed the enactments that had borne most heavily on them; and asked their aid in repealing certain other Statutes that had imposed civil disabilities on them, and on the Papists. The Nonconformists saw the snare laid for them, and they refused to enter it. They declined to recognise a community of standing between the Papists and themselves. They knew that they had been excluded because of Truth: the Papists because of evil. Rather than purchase privileges for evil, they preferred to remain unprivileged themselves. They did not think that the principles of civil and religious liberty required that no distinction should be made between superstition and Truth; blasphemy, and the acknowledgment of God. Because men, for their own purposes, may find it expedient to ignore the line that separates Truth and Falsehood, they would do nothing that would imply that *no* such line existed. They would not strike the sword out of the Ruler's hand because of the possibility that the Ruler might use it amiss. They would rather trust God respecting that, and allow the sword to remain where God had placed it.

All to whom God has given positions of influence (whether as parents, masters, magistrates, legislators, or electors) are responsible to God for the exercise of that influence. They are responsible for favouring and honouring that which is good, and for refusing to countenance that which is evil. What if a parent should resolve to smile on all his children alike, and were to make no difference between profanity and virtue, lying and truth? Or what if the child who was most morally corrupt, should, because of his attractiveness, or the superiority of his intellectual powers, be favoured most? Should we expect such a family to prosper? Should we not anticipate a curse? It is not otherwise in a State. The Scripture tells us what to expect when "vilenesses" [זֻלּוּת] are exalted. (See Psalm xii. 9.)

We hear much said respecting "the rights of men." But has not God also *His* rights. If men turn from Him and worship that which their own hands have made, if they refuse to acknowledge the supreme and sole authority of Holy Scripture, if they insult His name, or deny His existence, they infringe the rights of God; and they who favour, or honour, or promote those who do such things, become partakers of other men's sins.

I extract the following from the Baptists' Confession of Faith, presented by them to Charles II.,

March 1660; "for which (say they) we are not only resolved to suffer persecution to the loss of our goods, but also life itself, rather than decline from the same."

"We believe that the same Lord Jesus, who
" showed Himself alive after His passion by many
" infallible proofs (Acts i. 3), which was taken up
" from His disciples and carried up into heaven
" (Luke xxiv. 51), shall so come in like manner
" as He was seen go into heaven. (Acts i. 9, 10,
" 11.) 'And when Christ who is our life shall ap-
" pear, we also shall appear with Him in glory.'
" (Col. iii. 4.) For then shall He be 'King of
" kings, and Lord of lords.' (Rev. xix. 16.) 'For
" the Kingdom is His, and He is the Governor
" among the nations' (Ps. xxii. 28), and 'King
" over all the earth' (Zech. xiv. 9), and 'we shall
" reign with Him on the earth.' (Rev. v. 10.) The
" kingdoms of this world (which men so mightily
" strive after here to enjoy) shall become the king-
" doms of our Lord and His Christ. (Rev. xi. 15.)
"'For all is yours (ye that overcome the world),
" for ye are Christ's, and Christ is God's.' (1 Cor. iii.
" 22, 23.) 'For unto the saints shall be given the
" kingdom, and the greatness of the kingdom, under
" [mark that] the whole heaven.' (Dan. vii. 27.)
" Though (alas!) now many men be scarce content
" that the saints should have so much as a being
" among them; but when Christ shall appear, then
" shall be their day, then shall be given unto them

"power over the nations, to rule them with a rod of iron. (Rev. ii. 26, 27.) Then shall they receive a crown of life, which no man shall take from them, nor they be by any means turned or overturned from it, for the oppressor shall be broken in pieces (Ps. lxxii. 4), and their vain rejoicings turned into mourning and bitter lamentations, as it is written" (Job xx. 5—7).*

Suppose that the 20,000 men who "owned and approved this Confession had been asked to unite with Papists, and Anglo-Catholics, and Infidels, in an attempt "to separate civil duty from religious Truth," what would have been their reply? They would have said, "We choose rather persecution, imprisonment, and death. Through the Lord's help we will struggle on to our life's end *against* Falsehood, and *for* Truth; but we will fight only with spiritual weapons, and we will ally ourselves with no Ahabs. We remember what befell Jehosaphat. We desire not that that which was said to him

---

* This confession is "subscribed by certain elders, deacons, and brethren, met in London, in the behalf of themselves and many others unto whom they belong, in London, and in several counties of this nation, who are of the same faith with us." Then follow forty-one names, after which is written, "Owned and approved by more than twenty thousand." Hence it appears that the Dissenters had once the honour of contending for the personal reign of Christ on earth, and of suffering for professing of the same.—*From Crosby's History of the Baptists, Vol.* ii. *Appendix, p.* 85.

should be said to us: 'Shouldest thou HELP
THE UNGODLY, and love them that hate the Lord?
Therefore is wrath upon thee from the Lord.'"
(2 Chron. xix. 2.)

It may safely, I believe, be affirmed that such
would have been the language of those who
signed this Confession. They had not despised
the teaching of Daniel and the Prophets. But it
must be acknowledged that the Nonconformist
body as a whole, were not prepared to follow in
this path. If they had been, their history would
not have been what it is. The Political Dissent of
modern days would have had no existence.

The stedfast resistance of the early Nonconformists to the doctrines and order both of Popery
and Anglican Sacerdotalism, cannot be too highly
praised. Nevertheless even they were not careful
to use spiritual weapons *only;* nor did they keep
themselves separate from men who knew not God.
Their principles too were defective. They virtually
accepted as their motto, *Vox populi, vox Dei.* They
favoured Democracy both in the Church and in the
world. They seemed to imagine that through them
the truths of the Gospel would penetrate Society, and
that if Governments could be made the expression
of the mind of Society after Society had been thus
penetrated, mankind would be regenerated, and the
sovereignty of the world become the sovereignty of
our Lord and His Christ. What the Scripture reveals respecting the course and end of *this present*

age (αἰωνος) they considered not; nor did they recognise the future of converted Israel and of the nations, in "*the age to come.*" "The times of the Gentiles,"—that is to say, the whole present period of Gentile supremacy (a period which has at its commencement *Nebuchadnezzar*, at its centre *Pilate*, and at its end *Antichrist*) is the period which they looked on as including within its scope THE MILLENNIUM!! They imagined that the Gospel had long commenced to smite the Image of Gentile greatness, and was grinding it to powder. Their thoughts on these subjects seem to have coincided with those of Lightfoot. These were fatal errors, seldom, if ever, escaped by those who neglect prophetic Truth.

"Lex rex" is the compendious form which men have chosen to express the conclusion at which they have arrived respecting Government. But it embodies a sentiment which God has not sanctioned. In Heaven a *Person* is recognised as King: and that Person is the One Living and True God. None speak in Heaven of "Law" being "King." "Law" is recognised as proceeding from the King, and is known to be perfect because He is perfect. This is the order of Heaven, and in the Millennium it will become the order of earth. But it is otherwise now. Men have discovered to their sorrow that no one person among themselves can safely be entrusted with the origination and ministration of Law. Various devices have been, in past ages, tried, to make power in the hands of men, bearable; but they have failed. The last device is to sink the Ruler into a mere functionary, and to exalt "Law" (that is to say, the will of Society—the will of the governed) into the Throne. Men have long tried this plan; but they are not yet satisfied. There are conflicting opinions, and conflicting interests which they know not how to reconcile. They imagine now that religious truth, or that which professes to be religious Truth, is the great cause of discord. Instead of waiting to ascertain whether or not Truth can be distinguished from its counterfeits, their resolve is to regard all religious truth as a sword in the household, and to cast it out. Let men's laws, say they, be formed without reference to God's laws, and so governmental harmony may be reached. In Paris in 1792 this plan was thoroughly tried. What were the results? Licentiousness, bloodshed, and misery, until a despot came, and was hailed as a deliverer.

In the second and third decades of the present century, the results of the convulsions of preceding years developed themselves in England. Democratic liberalism strengthened itself greatly during the revolutionary wars. Roman Catholicism saw its opportunity, as in the days of James II., and courted the Liberals. Its advances were not repelled. The vigour of Protestantism had waned. It had become, throughout a wide sphere, nominal, educational Protestantism—Protestantism that cared more for its own social status than for Truth. An alliance between the Catholics and Liberals was formed. The Catholics were emancipated. The Liberals were recompensed by the Reform Bill. The Irish Church was assailed, and its dioceses summarily re-organised without its concurrence being granted, or even requested. The Established Church was startled, and resolved, under new leaders whom Oxford supplied, to abandon Hooker, and to retire upon Laud and Rome. Two Schools arose in Oxford—a sacerdotal School which founded itself on a non-natural interpretation of the Articles; and a Rationalistic School which founded itself on a non-natural interpretation of the Scripture. Revealed Truth became a plaything. The Word of God was made of none effect. If used, it was used to blind and to deceive.

Darkness came in, and strong delusion. Yet at the same moment God mercifully interfered, and on behalf of those who feared His name, drew aside the veil that had for centuries been cast over Prophetic Scripture. Isaiah, Daniel, and the Revelation were read and understood. The coming period of Antichristian Apostasy, and the glory that is to follow, the place of Israel in that Apostasy and their subsequent place in the reign of glory; the place of the servants of Christ in suffering now and in glory above the Heavens when their Lord shall have come—these, and other like things, were apprehended with a distinctness and power that caused many to feel that they were enabled (in a manner unknown before) to understand and use the Scriptures *as a whole*.

But the Nonconformists for the most part, preoccupied with other things, discerned not this interposition of the hand of God, and refused the light that had been given. They might have rallied to the Bible. They might have repelled the advances of the Romanists, and have avowedly raised an exclusively evangelical banner. They might have placed themselves within the circle of revealed Truth, and declined to fight with any except spiritual weapons. They might have said, "We will unite with none except those who accept the Bible in its integrity, and recognise it as the one only Rule." God would have blessed them then: the advancing floods would in all probability have been stayed. But they determined otherwise. They united politically with the foes of the Bible, and fought by the side of those who were fighting against Protestantism and Bible-Truth.

The results are manifest. They could scarcely be defined more clearly than in the words of Mr. Gladstone already quoted. The Protestant character of the Legislature was first destroyed—next its Christian character, and now its Theistic. A person may now be a legislator who denies the existence of God. The principle that "civil duty is to be separated from religious Truth" is now established, and all this is considered to be a triumph!! Apart from the Nonconformists, these steps would not have been taken: these triumphs would not have been gained.*

I cannot but hope that there are many among the Nonconformists who have anxious misgivings as to the rightness of the path pursued. May they have grace to consider their ways, and to make open and unreserved confession. They ought not to have allied themselves with the enemies of God's Truth. They might have distinguished between the abstract truth of certain principles and their misapplication or perversion, and have, whilst assailing the perversions, have vindicated the principles. In struggling against *unrighteous* tests and restrictions they need not have destroyed *righteous* tests and restrictions. Probably if left to themselves they would not have done this. But they linked themselves to aliens. They had disobeyed the explicit command, "Be ye not unequally yoked together with unbelievers. . . . . What concord hath Christ with Belial?" The consequences of their disobedience in these things, and their neglect of

---

* Mr. Bradlaugh, addressing his constituents in the Town Hall, Northampton, is reported to have said :—" He would not go over the past struggle, but he could not fail to recognise that it was by the solid vote of the Nonconformist party, scarcely broken, that he was put in the position he had sought. The statement was received with cheers." (See "Daily Telegraph," October 29, 1880.)

the light which Prophetic Scripture sheds upon the future, have been disastrous beyond conception. Eternity alone will reveal the real character of the results, and the extent of the ruin wrought. "The narrow ledge of Theism," as it has been called, was infinitely important in God's sight; and if the Christian Nonconformists of England, instead of leading the assault, had aided in the defence, "the narrow ledge" would not have been abandoned. But they did otherwise. They joined the enemies of the Bible, and accepted the principle that civil duty is to be separated from religious Truth. That which has been done cannot be undone. The flood-gates have been opened, and they cannot now be closed. Fifty years ago it would have been, I believe, quite possible to have required and obtained from every Legislator an acknowledgment of the existence of a God, of the Lordship of Christ, of the inspiration of Scripture, and of its supreme and sole authority as a rule. Such tests (though not all that could be desired) would have preserved this country from becoming governmentally Atheistic, Deistic, Romanist, or Pagan. But the opportunity has been lost, and will never again recur.

Will the Nonconformists recognise the greatness of the sin that has been committed, and openly confess it before God, and before men? Will they separate themselves from aliens, and return to Scripture and to God? If not, what can they expect? Can they expect to be kept from that "hour of temptation which is coming on the whole world [ἐπι της οἰκουμενης ὁλης] to try them that dwell upon the earth"? I believe not.

It is no joy to me to write these things. I write them with deep sorrow. I write them not as an enemy, but as a friend. I see the pitfall, and I cannot but warn of it, however much my feeble warning may be spurned, and myself despised.

According to present appearances the political Nonconformists will, during the present year, have an opportunity once more afforded them of either manifesting their repentance, or of persevering in the path which they have so long trodden. They will, no doubt, be asked to aid those who wish to place Protestantism under the feet of the Papacy in Ireland. They will be asked also to unite with those who (not because they oppose error, but because they hate Protestantism and the Bible and Truth) are about to assail the establishment in this country. They will be urged to sustain the principle, that civil duty is to be separated from religious truth. If because of party-interests and social advantages they consent to follow in the paths thus opened, if they refuse to humble themselves and to consider their ways, and to turn to Him from whom they have greatly revolted, I see nothing before them but ruin. "Behold, all ye that kindle a fire, that compass yourselves about with sparks: walk in the light of your fire, and in sparks that ye have kindled. This shall ye have at my hand; ye shall lie down in sorrow."

There will, I trust, be individuals who will read aright the signs of the times, and will break the links that unite them to the onward progress of godlessness. But let them take heed that they form no new links rashly. Let them gather to none among whom the Bible does not rule. Many besides Papists can speak of "the one Body," and of "the living voice of the Spirit in the Body," and of the Church "dowered" with grace and gifts by a heavenly Bridegroom. Let them remember that grace and gifts are not found where the Bible does not rule. He that "abideth not in the doctrine of Christ, hath not God." The doctrine of Christ can be received only through His written Word. In the days of Israel's calamity which Jeremiah witnessed, it was long before Ezra and Nehemiah came. But they came at last: and the Scriptures were again read and understood, and a remnant, though they were few and feeble, heard and responded to the words, "Let joy in Jehovah be your strength."

APPENDIX.

ANTICHRIST TO BE KING OF ASSYRIA AND BABYLON—PRESENT PROSPECTS OF THOSE COUNTRIES AS SHOWN BY THE OFFICIAL REPORT OF GENERAL CHESNEY.

ANTICHRIST is several times referred to in the Scriptures as "King of Babylon" (Isaiah xiv. 4)—and "King of Assyria" (Isaiah x. 12). He is also called in Daniel "King of the North," *i.e.* King of Syria. Syria, as so applied, includes Assyria, and all the Euphratean district.

In the foregoing pages, I have not entered into the question of Antichrist's relation to Assyria and Babylon, because I have elsewhere considered it. In the book advertised at end and entitled, "Babylon—Its Revival and Final Desolation," I have shown from Scripture that Babylon, at the period of the Lord's return, will have recovered a condition of prosperity exceeding even its former greatness. Like Jerusalem, Egypt, and other countries of the East, it will revive for a season under the hand of man. The blow which brings it into its final state of utter desolation, is to be inflicted when Israel is forgiven, and when the Day of the Lord comes. I have also shown, on the authority of General Chesney and others, that the present condition of Babylon and Babylonia, is not one of *utter* desolation, such as the Prophets describe. The site of Babylon has always been inhabited, and is, at this moment, the seat of a considerable city containing probably 10,000 inhabitants.

The thought of Antichrist's connexion with Babylon was preserved in the early ages of Christianity: and has not been entirely quenched even in its darkest periods. *Cyprian*, in the third century, applies to Antichrist that remarkable passage in Isaiah which describes the greatness and fall of the last King of Babylon—"Is this the man that made the earth to tremble, that did shake kingdoms?" (Isaiah xiv. 16.) *Jerome*, in the fourth century, speaks of Antichrist as about to arise from among the Jews, and *to come from Babylon*. [Nasciturus de populo Judæorum et de Babylone venturus—*Jerome on Dan.* ii.] *Cassiodorus*, in the sixth century, again suggests the thought of Babylon meaning the Chaldean city.* *Aretas* does the same. *Bede* also, in the eighth century, in his remarks on the Apocalypse, hints at the possibility of the Euphratean City being the place of Anti-

---

* See "Prospects of the Ten Kingdoms," p. 393.

christ's greatness. In the twelfth century, *Roger of Hoveden* says: " in the City of Babylon which was of old the illustrious and glorious city of the Gentiles and the head of the kingdom of the Persians, Antichrist will be born; after being born in the kingdom of Babylon, he will come to Jerusalem." (Scrip. Anglicani, p. 681.)

Even as late as the sixteenth century, we find *Malvenda* in his elaborate treatise on Antichrist, speaking of *Babylon* as the place from which he is to arise. He connects with this thought the fact of the Jews (from whom he expects Antichrist to spring) having been so long established in the Euphratean regions. He quotes Josephus in proof that two cities of Babylonia (Nearda and Nisibis) had been for a long time the chief depositories of Jewish wealth. Thence also the Jews promulgated their Talmud in the latter part of the fifth century. "Although," says Malvenda, "the celebrity of these Talmudic Schools did not continue after A.D. 1024, yet it cannot be denied that innumerable Jews are up to the present moment (*i.e.*, the sixteenth century) collocated in those regions, and that it still continues the great conservatory of Jewish doctrine." Malvenda supposes that the long settlement of the Jews in Babylonia, will make it more easy for Antichrist, as thence arising, to pretend that he is descended from their royal line. (P. 82.)

"All," continues Malvenda, "are aware that it

"is a universally-admitted truth,* that Antichrist is to be born in the city of Babylon. Jerome, in his commentary on Daniel ii., expressing his own judgment, and the judgment of all the Fathers of the Church, says: 'Our writers interpret all these things of Antichrist, who is to arise from the people of the Jews, and to come from Babylon.' Bede, referring to this opinion of Jerome and of the Fathers, observes, when commenting on the seventeenth of the Revelation: 'Some interpreters say that Antichrist, having his origin from Babylon, will overthrow the King of Egypt, Africa, and Ethiopia.' Aretas, when explaining the following passage in the Apocalypse—'Loose the four angels, who have been bound in the great river Euphrates,' says: 'the thought is not be rejected that dæmons are bound at the Euphrates, since in a short time Antichrist is to proceed thence, arising from among the Hebrews who are in captivity, either those reserved in Jerusalem, or those who have settled in those (Euphratean) districts."

"The reasons for which it hath pleased God to appoint that Antichrist should arise from Babylon rather than elsewhere, He hath buried deep in the secrets of His own bosom: nor has He, up to the present time, revealed it, as far as can

---

* "Communi omnium consensu receptissimum esse norunt omnes, Antichristum in Babylone urbe in lucem edendum."

" be ascertained, to any among men. Neverthe-
" less the erudite and pious meditation of some
" orthodox believers has alleged * * * * the fol-
" lowing fitting reasons for the Divine decree. In
" the first place, as Nimrod, the founder of Babel,
" *i.e.* the Tower of Babylon, a savage Tyrant and
" cruel oppressor of men, was the person who de-
" clared open war against God; so it is meet that
" there should arise from the self-same Babylon,
" the last and most atrocious persecutor of the
" Church—Antichrist. Moreover, seeing that Nebu-
" chadnezzar and Antiochus Epiphanes—two Mo-
" narchs who bore upon the Church with an over-
" whelming power of destruction, and who were the
" Antichrists of the Old Testament, and remark-
" able types of the Antichrist that is to come—
" seeing, I say, that these Monarchs reigned in
" Babylon—it is fitting that the true Antichrist of
" the New Testament should arise from the same
" Babylon."

" Besides, no place can be pointed out more
" meet for the nativity of Antichrist than Baby-
" lon, for it is the City of the Devil; always dia-
" metrically opposed to Jerusalem, which is deemed
" the City of God: the former City, *i.e.* Babylon,
" being the Mother and disseminator of every kind of
" confusion, idolatry, impiety; a vast sink of every
" foul pollution, crime, and iniquity; the first city
" in the world which cut itself off from the worship
" of the true God; which reared the citadel of uni-

"versal vice; which perpetually (according to the "record of Holy Writ) carries on the mystery of "iniquity, and bears imprinted on her brow the "inscription of blasphemy against the name of "God. The consummation therefore of impious-"ness, which is to have its recapitulation in Anti-"christ, could not break forth from a place more "fitting than Babylon."

I make these quotations, not as agreeing with every sentiment they express — much less with the doctrines of the writers on other subjects: I have given them merely to show, that the thought not only of the personality of Antichrist, but of his connexion with Babylon, has never been entirely extinguished, even by the black corruptions of Romanism. It seems to be reserved for the worldliness of Protestantism to quench this remaining light, and then to cheer men onward in the course which is to end in establishing "in the Land of Shinar" (see Zech. iv.) that system of "wickedness," under which Babylon, Israel, and the nations will ripen for their final doom.

There is no joy, no gratification in being obliged to look forth upon the rising prospects of the nations—prospects, which are interesting the hearts and employing the energies of millions, and to say that they will soon end in overwhelming ruin—to say that the influences now being developed around us are urging men onward to the point where they will be confronted by the stern but righteous judgments of God. But we dare not say otherwise; we dare not hide the Scriptures of God, nor prophesy peace where there is no peace.

This our Protestant country has taken the lead in constructing that new governmental system which avowedly excludes a regard to God's truth in legislation and government: and calls together into Satanic union, Mahomedanism, Judaism, Infidelity, and false Christianity, that they may together fashion educationally, morally, and religiously, the hearts of men. This country has taken the lead in indissolubly connecting these principles with that rising system of commercial greatness by which she expects the nations to be swayed; and intends that wealth, derived from commerce so pursued, should be the pillar of the social system. This country also has been the first to enter those once-flourishing regions of the East, which have been smitten under the judgments of God, and are still obnoxious to His curse: it has entered those regions with the view of considering how far they can be recovered from their ruin, and again made the depositories of wealth and civilisation.

Some years ago, the Government of this country sent out an expedition to survey the Euphrates, and report upon the facilities it afforded for commerce, and for becoming the channel of communication between Europe and the East. The expedition was commanded by Colonel Chesney, and the Report which he has just published, affords abundant evidence to the zeal and ability with which he fulfilled the mission.* Few after reading it will question the capabilities of those regions. Can it be doubted that the tempting picture will act upon the cupidity of the present age? Is it likely that they, whose energies are exploring every land and traversing every sea, will forbear from entering rivers such as the Euphrates and Tigris, situate as they are in the midst of the ancient garden of the earth, and having beyond them the Indian seas, which have now become the seat of European civilisation? They will certainly be entered, and probably very soon. God indeed only knows the time, but He has said that "the Ephah," with "wickedness" hidden in its midst, is to be established in the Land of Shinar, and His words must be accomplished in their season. In the meanwhile, whilst we behold the latitudinarian infidelity of the rising commercial system strengthening around us, and if we should live to see the river of Babylon re-entered and traversed by successful energy, and Babylon and the Land of Israel touched by the hand of Western civilisation, and revived, may we then remember the words of the angel who cried, saying, "Fallen, fallen, is Babylon the great, who

* It is entitled, "The Expedition for the Survey of the Rivers Euphrates and Tigris, carried on by order of the British Government in the years 1835, 1836, and 1837, by Lieut.-Colonel Chesney, R.A., F.R.S., &c., Commander of the Expedition." (Longmans. 1850.)

made all nations drink of the wine of the wrath of her fornication :"—may we more anxiously than ever read the circumstances of that day of man's chiefest glory, in the light of the closing testimonies of the Book of God.

The following quotations from General Chesney's work will sufficiently show the opening prospects of those regions :

"The river now about to be described (*i.e.* the Euphrates) rises at no great distance from the shores of the Euxine, and in its course to the Indian Ocean, almost skirts those of the Mediterranean. * * * * The Euphrates at one time formed the principal link connecting Europe commercially with the East. Its historical celebrity has excited in its favour an interest superior to that which has been felt for any other river; and it may reasonably be expected, that when its advantages shall be fully known, and duly appreciated, it will rise to a high degree of *political* and *commercial* importance."

" In a range of more than 1,780 miles from its eastern source, this river may be said to unite three great and important seas ; which, without it, would be destitute of any water communication with each other, whilst the varied productions of the intervening territory would, in a great measure, be lost to the rest of the world." (Vol. I., p. 40.)

"*Bir* is one of the most frequented of all the passages into Mesopotamia, and about sixteen large passage boats are kept, * * * for the use of the caravans, which occasionally number 5,000 camels." (P. 46.)

"This great river then proceeds through the Date-groves * * * across a bare country onwards to Hillah. * * * This Town *is built on a part of Babylon, and chiefly with materials obtained from its ruins:* it contained, in 1831, the time of my first visit, about 10,000 inhabitants, whose dwellings are principally on the right bank; the line of houses forming an obtuse angle, almost midway between the Mujellebé and the still more celebrated Birs Nimroud."* (P. 57.)

Extracts of Letters to Colonel Chesney, from Officers sent by him to explore the capabilities of the Euphrates for Steam Navigation and Traffic.

"SIR,

"The noble and interesting river Euphrates is far too celebrated to require from me more than a fair view of the prospect it offers for establishing an economical and more rapid communication between Great Britain and her Indian possessions, than has hitherto been attained. The brilliant prospects of a new channel being opened to our enterprising mercantile world through a steam establishment on the Euphrates, ought to awaken our best energies."

"The Town of Hit produces salt and bitumen to any extent:—190 miles below Hit is Hillah, a very considerable Town, having a bridge of boats across the river, which being under the control of the Musellim of the place, can always be opened by his orders on the approach of steamers."

"Basrah is about forty miles below Kurnah, and I consider it admirably suited for the magazines, dockyards, etc., of a large force."

"I consider that a rapid steam voyage may be performed, both up and down the Euphrates, at any season of the year." (P. 687.)

(Signed) "R. F. CLEVELAND, R.N.
"Dated 17th July, 1836."

* For a further account of Hillah and the condition of Babylon itself, see "Second Series of Aids to Prophetic Enquiry." For many of the particulars there given, I was indebted to the kindness of General Chesney.

Extract of Letter from E. P. Charlwood, Esq., R.N., to Colonel Chesney. (P. 691.)

"The Arabs always evinced great eagerness to barter their provisions, and in fact every thing they possessed, for our Glasgow merchandise, * * * * so that I am convinced, considerable commerce would be carried on with great success on the river. Taking all these things into consideration, I should say it would be highly advisable to navigate this river, as being the *speediest* and most secure route between Great Britain and her Indian possessions."

Extract of Letter from James Fitzjames, Esq., R.N., to Colonel Chesney. (P. 694.)

"The advantages that would ensue from the establishment of a regular steam communication on the Euphrates, would, I am convinced, amply repay any outlay and trouble which might attend the commencement. The avidity with which the inhabitants of the different towns on the river bought our Manchester woollen goods, etc., sufficiently proves, that a great opening is presented to our commerce. Aleppo, Bagdad, Basrah, and (should the Karim be navigated) Ispahan, would soon become marts for British produce, and the influence of the British name be thus increased and extended."

"Taking these things into consideration, it appears to me, that England would not have cause to regret having made the Euphrates the high road to her Indian possessions, even should it afterwards be found that letters and passengers might be conveyed with more speed by the line of the Red Sea."

"A splendid road might be made over the 100 miles which separate the Euphrates from the Mediterranean. I should think a railroad impracticable, but I think a canal might be cut. * * * This would complete the communication by water, of England with India, by the shortest possible line."

Extract from Letter of W. Ainsworth, Esq., Surgeon and Geologist to the Expedition.

"The river Euphrates is evidently a navigable stream. I am acquainted with it * * * from the Taurus, to its embushure in the Persian Gulph, a distance of upwards of 1,200 miles; and in that extent, there are only two real difficulties, both of which are superable, by undergoing an expense quite disproportioned to the importance of rendering efficient at all seasons of the year, and throughout so lengthened a course, the navigation of this noble river. * * * In a commercial point of view, the close communication thus established with so great an emporium of trade as Bagdad, is of the very first importance; nor is the connexion that would be established between Basrah and Bagdad of a trifling character; but there are also on the river between Kurnah and Felujah, large towns, as Sheikhel-Shuyakh and Hillah, and powerful tribes, as the Mountefik Arabs, who have long been actuated by the spirit of commercial enterprise."

"There is, indeed, amongst almost all the tribes a cupidity that is easily aroused, and which would stir up the people to new exertion, in order to obtain comforts and luxuries with which they would then first become acquainted, and would not be slow in appreciating. The boasted frugality and indifference of the Arab, are not proof against the inventions of an improved mechanism in cutlery or fire-arms; and nowhere is there displayed a greater anxiety for gay dresses and ornaments: this taste has become almost a passion with both sexes. We have abundant evidences of the love of decorating their children, and of a desire to improve their condition."

"The advantages which are presented by the opening of the navigation of the river Euphrates, belong to universal civilisation, as well as to increase of national power. The waters of this great river flow past the habitations of four millions of human beings, amongst whom their own traditions have transmitted the sense of a revolution to be effected by the introduction of a religion of humility, of charity, and of forbearance."

"The intellectual powers of the descendants from the most noble stocks of the human race, are not extinct in their present fallen representatives, and it would be difficult to say to what extent civilisation might flourish, when revived in its most antique home."

"The national importance of this navigation, is of the most comprehensive character. All acquainted with the history of the communication of nations, which, as Montesquieu has ably pointed out, is the history of commerce, must be aware, that those circumstances which led to the annihilation of the commerce of the East, would be revolutionised by the opening now proposed to be effected; and that whilst civilisation might be confidently expected to return to its almost primeval seat, it would do so under a very different aspect, and with vastly improved means, over the days of Opis and Ophir, or of Caucasium and Callinicum." "All these advantages are to be obtained by the navigation which you have entered upon, and of which you have proved the practicability." (P. 697.)

Sir A. Layard, writing to an eminent English merchant in 1843, says, "I believe Susiana to be a province highly capable of the most varied cultivation; the soil is rich, labour cheap, the inhabitants well disposed, and the country traversed by several noble rivers: * * * the land is highly favourable for the cultivation of *cotton*, which is now much neglected, but which might be much improved. I made many enquiries as to the growth of hemp, * * * and I found the country well adapted for its cultivation." (P. 701.)

"At Mosul (Nineveh) a considerable opening for British commerce exists. The present consumption of British goods in Mosul and the adjacent country, is more than sufficient to support a mercantile establishment. * * * A piece of print worth thirteen shillings in Manchester, is sold in Mosul for thirty-two. * * * The trade of such an establishment would probably soon extend into Persia, where Russian trade is now increasing." (P. 703.)

"Notwithstanding all the existing disadvantages, boats with merchandise are continually tracked up the rivers in Mesopotamia; but the fleets going up the Tigris against the stream, from Basrah to Bagdad, consume from thirty to forty days, while a steamer would perform this distance in four days and a half." (P. 705.)

"Good freights are therefore secured for steamers, and a valuable opening presented for trade, since an Arab population of about *twelve millions* is to be supplied. The actual trade to Bagdad was in 1833—12,000 bales or packages, brought up the Tigris at a freight of £1 per bale."

"The establishment of the navigation, would probably lead to that of English mercantile houses at all the chief places of trade on the Euphrates and other rivers and branches at the interior stations." (Pp. 704, 705.)

"*Armenia* is the territory ranging along the course of the Euphrates: * * * in the table lands, the soil is rich; it is but partially cultivated, but almost every kind of vegetable production is to be found."

"The wheat and barley are particularly fine; nor is it very uncommon to have three successive crops of grain in some places. The gardens yield grapes in abundance, also oranges, peaches, nectarines, figs, apples, pomegranates, and other fruits. Honey, wax, manna, and gall-nuts, are exported from the more mountainous districts, where, especially eastward of Tarabusim, the finest timber is very abundant. The scenery here is at once beautiful and strikingly grand from various points of view, as the mountains are seen rising abruptly from the sea to an elevation of four or five thousand feet, their sides being covered with dense forests, composed of gigantic chesnut, beech, walnut, alder, poplar, willow, ash, maple, and box trees, with firs toward their summits, and a magnificent underwood of rhododendron, bay, and hazel, etc. * * * The less elevated grounds produce cotton, hemp, tobacco, and raw silk in abundance; besides precious stones, such as the turquoise, beryl, chrystal, pearl, and ruby. Besides the more valuable metals, gold and silver, Armenia abounds in copper, lead, iron, saltpetre, sulphur, bitumen, quarries of coal, marble, and jasper, with several mineral springs, which have been celebrated for many ages."

"The Armenians are exceedingly fond of foreign commerce and home trade, both of which are prosecuted with such success, that even the Jews are in many instances driven out of the field of competition. The Armenians have been described as *brave*, a quality however that has long passed from them. They are now a commercial and agricultural people; well clad, abundantly fed, and possessing sheep, cattle, and fine horses in abundance." (Pp. 95—99.)

"The exports of Mesopotamia are: wheat, barley, rice, and other grains, horses, pearls, coral, honey, dates, cotton, silk, tobacco, gall-nuts, wool, bitumen, naptha, saltpetre, salt, coarse coloured cottons, fine handkerchiefs, and other manufactures of a country enjoying advantages which will eventually make its commerce more important than that of Egypt." (P. 109.)

"The numerous towns along the Euphrates, and the extensive population, partly permanent, and partly nomadic, on the banks of that river, will ultimately require several stations; but for the present one should be at Hillah (Babylon), and another at Anah, and a third at Beles."

"Though the subject has only been considered relatively to the people in their present state, it should not be forgotten that Mesopotamia possesses as many advantages as, or perhaps more than, any other country in the world. Although greatly changed by the neglect of man, those portions which are still cultivated, as the country about Hillah

(Babylon), show that the region has all the fertility ascribed to it by Herodotus, who considered its productions as equal to one-third of those furnished by all Asia. Being equal to, and in many respects even superior to Egypt with regard to its position and its capabilities, the time need not be distant when the date-groves of the Euphrates may be interspersed with flourishing towns, surrounded with fields of the finest wheat, and the most productive plantations of indigo, cotton, and sugar-cane." (Vol. II., p. 603.)

Extracts of the same character might be almost indefinitely multiplied; but enough has been quoted to show that even if we judged merely from the present aspect of circumstances, we should expect the revival of the Euphratean countries at no distant period.

Professing Christianity in the days of Constantine, was called upon to pronounce on the character of the change that then took place in its condition, and it declared the change to be of God. The consequences of that fearful error are now seen around us, both in the East, and in the West—on every side. "Iniquity abounds." Professing Christianity will be tried once more. It will have to pronounce on the character of that rising system, which will spread civilisation throughout the countries of the East, and reign over the Kingdoms of the Roman World. It will pronounce that also to be of God; and will thus seal its own condemnation. It will declare the masterpiece of Satan to be the work of God.

The fair scene of prosperity soon to be displayed in Jerusalem and in Babylon, will only be rejoiced in by those who blot from their remembrance the testimonies of Jeremiah and the Revelation. The words of such a chapter as the eighteenth of Revelation pronounce so solemn a denunciation on "civilisation revived in its antique home," that the thought of such revival, is shrunk from with horror by all who remember what God has spoken. "I saw another angel coming down from Heaven, having great authority: and the earth was lightened with his glory. And he cried with a mighty voice, saying, 'Fallen, fallen, is Babylon the great, and is become a habitation of demons; and a hold of every unclean spirit, and a cage of every unclean and hateful bird. Because by reason of the wrath of her fornication all the nations have fallen, and the kings of the earth have committed fornication with her, and the merchants of the earth have waxed rich through the power of her delicacies.' And I heard another voice from heaven saying, 'Come out of her, my people, that ye may have no fellowship with her sins, and that ye receive not of her plagues—because her sins have been builded together unto Heaven, and God hath remembered her unrighteousness.'"

These are the words to be held fast by all who fear God, and reverence His Scripture. They no doubt place us in sorrowful isolation. The Revelation was given that it might isolate us: but how blessed to be isolated in the midst of such a scene! God hath not said in vain: "Blessed is he who readeth, and those who hear the words of the prophecy, and keep the things which are written therein: for the time is nigh."

EXTRACTS FROM THE WORKS OF DR. LIGHTFOOT.

BISHOP WORDSWORTH having frequently referred to the works of Dr. Lightfoot, I have thought it well to add a few more extracts to those already given. His writings exercised considerable influence at the time of the Commonwealth; and as respects the Millennial reign of the Lord Jesus, are much in harmony with those of Bishop Wordsworth.

Dr. Lightfoot was born in 1602, and took a leading part in the Westminster Assembly, where he was regarded as an Erastian. During the Commonwealth he was appointed Master of Catharine Hall, Cambridge, an appointment he retained after the Restoration. The establishment of Christian Magistracy was to Dr. Lightfoot the evidence that Satan was bound and the saints reigning!

"In the text it is said, 'Judgment was given unto them'; what can this mean, but power and authority to be magistrates and judges? 'Yes (say our mistakers); it means that the saints, at the day of judgment, shall sit upon seats with Christ, approving and applauding His judgment.'

And they misapply other Scriptures as much for the confirmation of this, as they do this to such a construction. And those are Matt. xix. 28, Luke xxii. 30, which speak one and the same thing: where Christ speaks not at all of the saints judging the world, in such a sense as they feign to themselves, but only the twelve apostles [and Judas, if you well observe the places, to be reckoned for one] judging the twelve tribes of Israel. And the meaning is but this,—that 'when Christ should come to reveal Himself, in His glorious appearing in vengeance, against Jerusalem and the Jewish nation,—the doctrine that they had preached, should condemn the twelve tribes, that had not believed it, as if they themselves sat on the thrones to judge and condemn them.' And so some of the ancients have, of old, well understood it, that it is not spoken of their persons, but of their doctrine, judging and condemning.

"And to the true sense, that I say the text speaketh, speaketh also that equally abused place, 1 Cor. vi. 2: 'Do ye not know, that the saints shall judge the world'; *i.e.*, 'know ye not, that there shall be a Christian magistracy? that Christians shall be kings and magistrates, to rule and judge the world?'

"And the very same sense speaketh Dan. vii. 18, 26, 27; from whence both my text and that passage of Paul are taken; 'know ye not (saith he), that the saints shall judge the world?' How should they know it? Why, plainly enough out of that place in Daniel, where in ver. 18, it is foretold, that 'the saints of the Most High should take the kingdom; and possess the kingdom for ever and ever.' And in verses 26, 27; 'The judgment shall sit (as in the text), and the kingdom and dominion, and the greatness of the kingdom under the whole heaven, should be given to the people of the saints of the Most High.'

"Two considerations will put the matter out of all question.

"I. That the word 'saints' means not strictly nor really sanctified, in opposition to men 'not really sanctified,' but it means 'Christians in general,' in opposition to 'heathens.' And so the Apostle himself clears it in the verse before that I cited: 'Dare any of you go to law before the unjust and not before the saints?' What is meant by the 'unjust' there? 'Heathens,' or 'infidels,' as he calls them, ver. 6. And then, what is meant by 'saints?' but 'Christians' in opposition to 'heathens.'

"II. Observe the tenor of the contents in Daniel, and that will illustrate the sense of these verses that I produced. He speaks before of the four heathen monarchies, the Babylonian, Mede-Persian Grecian, and Syro-Grecian, that had had the kingdom, and dominion, and rule in the world, and had tyrannised in the world, especially against the church that was then in being: but at last they should be destroyed, and upon their being destroyed, Christ should come and set up His kingdom through the world; and then the kingdom, and rule, and dominion in the world, should be put into the hands of saints or Christians, and they should rule and judge in the world, as those heathen monarchies had done all the time before.

"And thus you have the words unfolded to you, and I hope according to the meaning of the Holy Ghost.

"And now, my Lords and Gentlemen, you may see your own picture in the glass of the text; for you are of the number of those of whom it speaketh. In it you may see yourselves, imbenched, commissioned, and your work put into your hands.

"In the first clause, the institution of the function the ordaining of magistracy and judicature : 'I saw thrones set.' In the second, the commissionating of Christians unto that office and function, 'They sat upon them.' In the last, the end of this office, and the employment they are set upon in it, 'Judgment was given unto them.'

"'Thrones set': by whom? By Him that had been the great agent in the verse before, Christ, that had bound the devil and chained him up.

"'*They* sat upon them':—who? They that are the persons mentioned in the verse before. Men of the nations, undeceived from the delusions of Satan, and brought into the truth of the Gospel.

"'Judgment was given them': for what end?

For judgment's sake, that they might execute judgment and righteousness among the nations.

"And so I have my words fairly cut out before me; and the matter and the method of the text call upon me to speak unto these three things:—

"I. Of the institution of magistrates, as an ordinance of Christ.

"II. Of Christian magistracy, as a gospel-mercy.

"III. The great work, the all-in-all of magistracy, the execution of judgment." (Lightfoot's Works, Vol. vi. p. 259.)

Sermon-Note on Rev. xx. 1—3. "And yet, all these thousand years he [Satan] was at liberty as before. He encompassed the earth, and walked in it as aforetime; caused the ten bloody persecutions against the Christians; caused the Roman emperors and nations to wallow in that bloodiness and filthiness, as they did: and yet, because he could not deceive the world with blindness and heathenism, as he had done,—but that the gospel was now come in, and undeceived men, and brought light and life among them,—he is said to be bound, and cast into the bottomless pit, and shut up. It is as chains and imprisonment, and the bottomless pit and hell, to him, that he cannot do the mischief, and work the destruction, by the blindness of heathenism, that he had done before." (Lightfoot's Works, Vol. vii. p. 413.)

www.ingramcontent.com/pod-product-compliance
Lightning Source LLC
Chambersburg PA
CBHW051852300426
44117CB00006B/362